The International Design Yearbook 2001

Based on an original idea by Stuart Durant

All commentaries and chapter introductions by
Jennifer Hudson unless otherwise indicated

Senior Editor: Paul Harron
Assistant Editor: Helen McFarland

Published 2001 by Laurence King Publishing
an imprint of Calmann & King Ltd

71 Great Russell Street
London WC1B 3BP
Tel: +44 20 7430 8850
Fax: +44 20 7430 8880
E-mail: enquiries@calmann-king.co.uk
www.laurence-king.com

Copyright © 2001 Calmann & King Ltd

All rights reserved. No part of this publication may
be reproduced or transmitted in any form or by
any means, electronic or mechanical, including
photocopy, recording or any information storage
and retrieval system, without permission in
writing from the publisher.

A catalogue record for this book is available from
the British Library.

ISBN 1 85669 236 1

Printed in Hong Kong

The International Design Yearbook 2001

Editor: Michele De Lucchi
Introduction: Jeremy Myerson
General Editor: Jennifer Hudson
Design: Lovegrove Associates

LAURENCE KING

Contents

Introductions	6
Furniture	14
Lighting	86
Tableware	126
Textiles	158
Products	178
Biographies	226
Acquisitions	234
Suppliers	238
Photographic credits	240

**Michele De Lucchi
Lamp prototype, The Lighting Field
1994**

Introduction

Jeremy Myerson

In the pantheon of modern Italian design, Michele De Lucchi is up there with the greats. He has often been described as the creative link between the pioneering post-1945 era and the contemporary design scene, a thread that runs from Ettore Sottsass and Achille Castiglioni through Andrea Branzi and Alessandro Mendini to De Lucchi's door. Yet he prefers to see his role as a standard-bearer from a broader transnational perspective. Taking a break during the long, arduous task of making his selection for *The International Design Yearbook 2001* in London, De Lucchi lifted his head from the light box, ran his fingers through his idiosyncratically long, scholastic beard, settled his spectacles on the end of his nose, and declared: 'My profession is not to be a designer or an architect – it is to make a bridge between industry and humanity'.

This role as mediator or interpreter between what he calls 'the cynical logic of the entrepreneur' and a world seeking greater meaning in the material objects all around it has interested De Lucchi for a very long time. It certainly influenced his selection for this publication. Any new idea regarded as capable of lifting the spirit or meeting a need was viewed positively. Anything deemed to be working against the human interest was destined for the reject pile, no matter how marketable it might be.

Fellow Italian designer Alessandro Mendini was once asked to give five reasons why he admired Michele De Lucchi. His succinct answers spoke volumes about De Lucchi's philosophy and achievements. First, Mendini admired the clarity of his ideas, right from the start of his career, and second, the independence of his formal language in the years after Memphis, the '80s movement that confirmed his international status. Third was De Lucchi's organizing ability, reflected in his successful design practice, and in his capacity to give younger designers a place in the team. Fourth, Mendini admired the ability to create forms that survive the transition from a small to large scale, so breaking down the barriers between design and architecture. Fifth, he admired De Lucchi's sense of poetry, which was so expertly applied to the design challenges of a domestic setting.

Clarity, independence, organization, scale and poetry are indeed the hallmarks of De Lucchi's career. One can see such virtues in the many aspects of his professional life. These qualities are also reflected back from the shimmering collection of furniture, lighting, tableware, textiles and products, the work of other designers, chosen so diligently by De Lucchi for this edition. There is a gentle tolerance, an acceptance of diversity, in De Lucchi's selection, indicative of a cross-cultural and inter-disciplinary characteristic that he admires in others, possesses in his own portfolio, and terms 'fluid thinking' (see p8). Yet his early involvement in the Radical Design movement of the early 1970s while studying architecture at the University of Florence (1969–75) was born of an impatience and intolerance with the status quo in Italian design at the time.

Radical Design was linked to student radicalism across European campuses at the time and set out to question mainstream capitalist ideologies, especially in city planning. The intention was serious, the techniques used were often experimental and surrealist, and based on pop culture. With three fellow students, Michele De Lucchi founded his own Radical Design group, Gruppa Cavart, in 1973 in Padua. Over the next three years Cavart produced manifestos, seminars, temporary structures and films about the future of architecture – material which contributed to the new Italian design of the 1980s.

In 1977, after working as a professorial assistant at the University of Florence and making his first forays into industrial design, De Lucchi moved to Milan, a defining moment in his career. He switched from architecture to design through work for Ettore Sottsass, Andrea Branzi and Gaetano Pesce. He began to contribute lighting to such groups as Studio Alchimia, influenced by his own research into 1950s Italian design. His bright, vibrant designs for hi-fi equipment, influenced by Op Art, appeared in the acclaimed architectural journal *Domus*.

Sleek, conformist modern design was by the end of the 1970s under sustained creative attack from a group of radical Milanese designers and architects. It was seen as aloof and elitist compared to the more democratic approach in Italy of the 1950s. As a new movement in design emerged with a new decade, De Lucchi found his milieu. Sottsass, a consultant to Olivetti since the 1950s, was central to his elevation to the main stage. In 1979 De Lucchi also began working for Olivetti and began collaborating with Sottsass on other projects. The intuitive ability of the older man to find ways to put abstract architectural forms into electronic products was exactly what De Lucchi, puzzling over how to reconcile the needs of production and people, was keen to explore.

When in 1981, Sottsass launched Memphis, it was inevitable that De Lucchi, then just 30 years old, would be a key lieutenant. De Lucchi was not just a creative spark within the group but also its tireless organizer, coordinating the work of its members and seeking financial backers. As Memphis shocked and thrilled in the early 1980s, De Lucchi's own objects displayed growing maturity and quality. Personal commissions from such firms as Artemide, Fontana Arte and Bieffeplast signalled the extent to which Memphis (and De Lucchi's role within it) had captured the imagination. But De Lucchi himself was keen to move on. In 1986 he launched his own provocative and radical Milanese group, Solid, to explore the idea of geometrics in design.

From tackling design from the viewpoint of an architect, De Lucchi returned to architecture and interior design – from the metaphor-based viewpoint of a product designer. At the head of his own growing practice from 1988 onwards, he went on to design more than 50 Fiorucci shops as well as a range of products and systems for Olivetti, mixing and merging the theoretical and practical disciplines of architecture and design to create a playful and poetic visual language which was very much his own.

Alongside commercial commissions and public manifestos, De Lucchi also initiated his own Produzione Privata in 1990 – a personal mechanism to make limited-edition objects in order to stimulate debate and initiate ideas away from commercial pressures. His interest in non-conformist cultural production, as opposed to mainstream market introduction, is well known and accounts for the many objects which result from cultural competitions and collectives shown in *The International Design Yearbook 2001*, from Hidden (pp34–5) to Droog Design (pp84–5). Produzione Privata was devised as a way in which handcraft could influence mass production, defining new tactile, formal and material qualities. As technology raced on, De Lucchi was anxious to protect and enhance the user experience, which gives any designed object or space its dignity and its meaning.

This humanist approach also made sense to some large commercial organizations whose faith in unbridled technological advance was becoming fragile by the early 1990s. In 1991, Michele De Lucchi won an international competition to design Deutsche Bank's branch offices. So began a period when his organizational skills and ability to bring on young designers would come to the fore. Although today De Lucchi has scaled back his practice to around 35 staff, he is designing the new generation of Italian post offices and his role as a design director capable of leading major programmes is widely acknowledged.

Unusually for such a high-profile design 'name', De Lucchi is very interested in the dynamics of the design team. His experience with Olivetti gave him a healthy respect for the achievements of in-house design, something which most stellar consultants tend to overlook. In the Products category of *The International Design Yearbook 2001*, a section which he describes as the strongest in the book, De Lucchi pays tribute to design teams at Siemens (p191), Philips (p185) and Sharp (p182). He regards Sharp's output as particularly consistent. The work of Jonathan Ive, design director at Apple Computer, also earns his praise. 'Only Apple could do what it's done,' he commented. What Apple has achieved is a new, much copied visual language for computer products, combining colour and translucency in components of supreme engineering precision and quality. The giant easel-like character of Apple's Cinema Display (p188) or the lightness of the Power G4 stack (p189) represent for De Lucchi near perfect examples of design as a bridge between industry and humanity. These are technological objects that demand your engagement and ownership.

As an architect De Lucchi has perhaps done his best work in product design, most notably on the famous Tolomeo lamp series for Artemide which unites technical precision with a poetic delicacy in updating the suspended-arm traditions of the Anglepoise and Tizio (see p13). So he is naturally encouraging of other architects venturing into furniture and lighting. Foster, Chipperfield, Moneo and Hadid all have work included in this edition. Norman Foster's lights for Artemide have even been specified by De Lucchi for the new Italian post offices. Moving seamlessly between design and architectural scale, from the small hand-held object to the vast public interior, is something with which De Lucchi has always been comfortable.

His architectural commissions in the early 1990s included a pavilion for the Groninger City Museum in the Dutch city of Grongingen, a crazily spectacular architectural amalgam composed by Alessandro Mendini and including contributions from Philippe Starck and Coop Himmelblau alongside De Lucchi's square brick building devoted to the city's history and archeology. On a project devoted to the most avant-garde instincts, here was the radical showing signs of restraint and an interest in tradition. Vittorio Magnago Lampugnani, former editor-in-chief of *Domus* has written of De Lucchi: 'Having been a design revolutionary ... he has turned into a mellowed advocate of a simplicity and discretion that is nevertheless not afraid to draw on the past. I find this development so interesting because in his case there is nothing revisionist, let alone reactionary about it.'

Quite how deep De Lucchi was willing to delve into history was revealed in a joint design project with Achille Castiglioni in 1990, an executive office furniture range called Sangirolama for Olivetti. This was based on the famous Antonelli painting of 1418, *San Gerolamo nello studio* (Saint Jerome in his Study). De Lucchi and Castiglioni were anticipating the softer influence of domestic furniture on the scientifically managed office. A decade on, that prediction has become a reality. De Lucchi is today designing offices for Armani using only home office furniture. The 'separation of the spheres', as historians have described the split 150 years ago between the home and the workplace, is no more. Work is re-entering the home at a rapid pace, driven by technology and changing business practices, and this trend is reflected in many of De Lucchi's furniture and lighting choices for this book. 'I expect a lot more innovation in the home,' he remarked during the selection.

What is clear is that complex, over-engineered office systems and products are inappropriate for working at home. De Lucchi also believes that simple reinventions of traditional objects are a necessary antidote to unsettling technological change, a point exemplified by his recent beech-and-leather benches for Poltrona Frau (see p8). In reconciling human needs with the demands of serial production, the timeless generic or classic in design has a key role to play in providing a starting point for the relationship with the user. This idea finds expression in De Lucchi's selection. Just look at Starck's Emeco chair (p32) or Emmanuel Babled's Murano glass (p153) or Christopher Deam's aluminium trailer (page 225).

New technological ideas are given due recognition in these pages, from smart textiles and LED (Light Emitting Diode) lighting to Ron Arad's computer-animated vases made using stereolithography. Yet there remains an admiration for simple, time-honoured things that work and reassure on a practical and emotional level for people. In 1986 Michele De Lucchi described his Solid design collection as 'above all an act of faith in the future; it exemplifies the fact that design, other than just produce beautiful things, can also generate practical and tangible progress'. Much the same could be said of his selection for *The International Design Yearbook 2001*.

Michele De Lucchi
Bench, Piazza di Spagna
Poltrona Frau, Italy
1999/2000

Introduction
Michele De Lucchi

I would like to express my sincere thanks to the publishers and to Jennifer Hudson for their kind invitation to be the Guest Editor of *The International Design Yearbook 2001*. It is an excellent opportunity for me not only to update myself about designs and production around the world, but also to check and reconsider my personal interpretation of the role of designers and architects (especially those closest to industrial projects) in the society and environment in which we live.

The exercise of choice for *The International Design Yearbook 2001* has been varied to say the least; I was expecting it, and wasn't disappointed. Quite the contrary. I therefore have grounds for thinking that presenting so much variety, energy and genuine enthusiasm will really be justified and useful. I viewed the 2,000-odd items sent in, which apart from a distinction by type of commodity, are impossible to catalogue in terms of style, character, figurative origin or iconographic reference. Linguistic limits and formal reference codes are increasingly hard to distinguish. I believe that, as I remember Herbert Schultes saying when he was in his last months as the director of design at Siemens, we have definitely entered into the era of 'fluid thinking', where everything must flow unimpeded, from culture to culture, discipline to discipline, role to role, skill to skill. We are in the first information and telecommunications technology era and are the first to enjoy this long-awaited worldwide orgy of connectivity, easy knowledge, abundance (already excessive) of information and images and richness of choice. We at *The International Design Yearbook 2001* team have learned something about richness of choice, having coped with the arrival of CD-Roms so crammed full of images that just looking at and analysing all of them seemed at times humanly too much to bear. And who knows what will happen in the years to come?!

Experts say that one of the abilities most in demand now and in the future is tolerance, and not just political and racial. I think that is true and that it is a trait required in our field as well, not so much in the sense of being able to accept things that are not pleasing and with which we do not agree but, above all, being able to put things together, combining and blending styles, visions and different technologies of opposite origins. Fluid, tolerant and creative: these are the premises for design, updated to the year 2001. The concept of creativity certainly could not be absent. How do you manage to hold together such disparate aspects and concepts unless you have a powerful stimulus towards research, innovation, new products, fantasy and dreams?!

At this point, this is my interpretation of the designer's role: to promote freedom of expression for everyone, using designs, creations, stories and metaphors. The designer and the architect are story-tellers, as Ettore Sottsass used to say in the 1970s during the time of Radical Design in Italy, and as Jean Nouvel has said in recent times at his wonderful conferences. A design is a story and the object – the product – visualizes the arguments of the narrative, be it in direct or symbolic form. Design is communication; it transmits an impetus of contemporaneity, involvement, of stimulus for freedom of expression. It is not that different from what we were saying, together with Ettore Sottsass, over 30 years ago when we claimed that the architect designs metaphors to stimulate the creative talents which are inherent and often hidden in the deepest recesses of everyone's personality. The difference is that this theory was at the time a provocative rejection of design of the traditional type; today it is a concept which helps us understand and accept what is happening in the world.

In my selection I have kept categories of similar products separate, as it has always been done in *The International Design Yearbook*. In 'fluid thinking', as well, the evolution of specific individual productive spheres remains independent. Furniture, lamps, accessories, fabrics, electronic products, vehicles, etc. logically go through different evolutionary phases, which can hardly be compared. The productive technologies, the characteristics of distribution systems and market demand, all still form a solid and concrete barrier to the homogenization of innovative movements. However, I have indicated similar linguistic categories to show emerging figurative trends wherever possible. It is still certainly interesting to compare chair with chair, table with table and lamp with lamp: it undoubtedly remains one of the major contributions which *The International Design Yearbook* offers with each annual edition.

I would, however, also like to stress a few points. The first concerns business initiatives: indeed, many events are linked directly to operations which are hard to define, initiated and promoted by individual, public or private investors who, apparently acting purely out of cultural interest, see in design a rich and fertile field. This applies to Hidden – Leon van Gerwen bought the sdb aluminium display company and called in a dozen designers with the proposal that they design products for the home uninfluenced by commercial pressures. The designers were able to propose furniture and objects for the home freely, without any brief, whether functional or market-driven. What emerged were products with a strong

Michele De Lucchi
Lamp, Rumi
From 'Thinking of Sufi Poets' exhibition
Produzioni Privata
2000

expressive charge, certainly new and in some cases provocative, just like the philosophy of this newborn company.

The 'provocative' approach also applies to the far more well-known case of Droog: also Dutch, it is a free, nonconformist group, united in the idea of renewing the meaning of design. Droog was founded in 1993 in Amsterdam and basically presents, not a catalogue of products, but a mindset to which to relate. Over the years, it has promoted and developed experimental projects to which young designers have contributed in particular. Droog's products have been marketed internationally up to now by the DMD company of Vaarburg. Droog is no longer concerned solely with products, but also with their presentation, display and the events which accompany them. In confirming that the interpretation of the disciplinary limits are changing rapidly, Droog is announcing that architects, interior designers and advertising and graphic agencies are now also collaborating in disseminating the success of Droog Design, that means dry design.

Symptomatic, entertaining, authentic, original and disruptive are words that certainly describe the catalogue *do*, in which no fewer than 10 designers or studios, including Martì Guixé of Spain, Thomas Bernstrand of Sweden, Radi of France and Dawn

Michele De Lucchi
Lamp, Nizami
From 'Thinking of Sufi Poets' exhibition
Produzioni Privata
2000

Finley of the United States have designed products for an experimental brand created by the Amsterdam advertising agency KesselsKramer. The title of the project, 'do create', alludes both to the designers and the consumers: the designer creates the products with which the consumer has to interact. The designer creates an unsigned product which is completed by the consumer, who adds his personal touch. The user is invited to intervene and play, he can influence the design and the product can identify the personality of its consumer. One no longer buys a form, a style or a function, now one buys an experience. Personally, I am a great admirer of Gijs Bakker and Renny Ramakers for the work they have achieved and the communicative force of their publications. I believe they represent the most innovative and enthusiastic design formula of these last few years, and I also admire them for having been able to sidestep traditional obstacles and difficulties in spreading an independent and original design culture.

Cappellini, albeit in certain more controlled and experimental movements, also every year surprises us by gathering around him excellent designers from all over the world and presenting projects and products that are always original, always fascinating, never banal; this year's selection is no exception.

And the public initiative of Kunstindustrimuseet and the exhibition for 1999 promoted by Louise Campbell, Cecilia Enevoldsen and Sebastian Holmbaeck called 'Walk the Plank' is notable. It is a friendly and unceremonious challenge, a search for ideas, imagination and skill, at a crossing-point between designers and cabinetmakers, between design talent and production talent. Some 20 furniture designers were invited to be paired together with as many cabinetmakers and each pair was given a plank out of which it had to make an item of furniture. The resulting pieces were exhibited at the Museum of Decorative Art in Gronnengarden and then sold at auction. The profit was then donated to a fund enabling furniture designers to experiment with new prototypes.

There are also many private initiatives by individual designers who, not having found sufficiently ample scope for research themselves in the industrial sector, create small but significant business initiatives which have had noteworthy success, often going well beyond the original aims. Ingo Maurer, Ron Arad and Philippe Starck have all experimented with the thrill of a business initiative based more on poetic logic than on a business model. It has often gone well, even very well as in the case of Ingo, who succeeded in turning a business into a wonderful poetical attitude. I myself can be catalogued in this sector with my Produzione Privata, whereby I try to simulate on a small scale what I would like to be able to do with the major industries, but which industrial logic finds impossible to accomplish. And it is precisely the experience with Produzione Privata which has helped me to understand this phenomenon and which leads me to talk about it now. It also drives me to underline the importance of crafts and the value of something still made 'by oneself', made by hand, in just a few editions, if not only one. Industry and crafts are not different when compared in terms of their contribution to the culture of design.

I have always upheld the role of crafts as a research forum for industry – an ideal forum, free from the constraints imposed by the size of the investment, the risk of research, the possible consequences of success or the lack of it. Working with crafts creates an ideal laboratory, the significance of which is not always understood in terms of its potential, and not always grasped in its current experimental role. I say 'ideal' because crafts combine technology and human talent, knowledge and sensitivity, know-how and skill. Crafts free the prototyping from any restraints and allow vast liberties of expression.

It is one of the prerogatives of the happy success of Italian design that despite the lack of new names it maintains its reputation as an international centre of design. In Italy, to the discredit of the industrial and political structure and to the credit of a limited number of businessmen, some extraordinary craft traditions have been kept intact which each year enable Italian and foreign companies to present at the Salone del Mobile [Furniture Fair] in Milan an infinity of new ideas, new models and new visions.

These craft traditions also encourage the phenomenon that tends to bring the world of design ever closer to the world of fashion; and this is to the absolute benefit of design because it can absorb from fashion its extraordinary inventiveness and productivity, the continuous capacity to generate innovation and interest, and to direct the world towards new and ever more appealing visions. Furthermore, it enables it, step by step, to witness the ongoing interesting alternation of choices, forms, ambitions and hopes. I would like this to be the most far-reaching role to be attributed to design.

I would also like to stress a point concerning design projects made by architects: I mean those who, in creating architecture, are obliged also to define the internal spaces and consequently often design the fittings for them. It involves a special way of seeing and understanding the design of objects for the home; I know this as a fact because here, too, I have direct experience since I am myself an architect. This selection has in fact been an opportunity for me to examine just how much architectural culture influences the industrial culture and how much, in my own work as a designer, derives from my original grounding. Clearly, architecture and design meet chiefly where one finds internal space and furniture, functional and aesthetic places and tools, environments which people inhabit and objects and utensils, whether fixed or mobile.

It is very difficult to describe concepts which depart from the meaning of words! It is equally difficult to understand why the worlds of spaces and things move, in terms of discipline, on two such diverse platforms. Only in rare cases do the distances fill up and the two worlds unite, amalgamate and combine, thereby enhancing one another. The architect designs the furniture and fittings according to his instinct of spatial definition, 'constructing' element upon element, like stone on stone, form on form, adding and subtracting, but nonetheless working in a frenzy of composition guided by his trade, a trade that has existed since the beginning of mankind. Fittings, furniture, objects, things, are today subjected to the industrial culture, which abides by completely different values linked to productive logic, technological convenience, market research and more generalized tastes. It is difficult to put together parameters so diverse and remote: only in a few cases has this been possible, precisely thanks to a third protagonist – unexpected and I really wonder whether always desired – namely, art. In certain cases, something greater than expected may happen – a wider concept which can govern one and the other and bring everything together in a single idiom. This is probably what we call art, but what I prefer to call the spirit of time, the spirit of all shapes, ideas, moods, the specific images of a precise historic moment. It is in this ideal place that architects and designers, with the ability to construct and produce, should meet. Furthermore, another challenging ideal of 'fluid thinking' is to combine architectonic culture and industrial culture, allowing the experience of a millenary culture to permeate the little more than centenary culture of an industry still struggling to show its best side, to display in the midst of the worst, whatever it is best at.

For this edition of the *International Design Yearbook* I have had the good fortune to put together the work of great architects like Foster, Sejima, Moneo and Chipperfield who with greater or lesser indulgence for industrial logic have produced significant furnishing projects. They certainly concede nothing to the enticements of mass produced products; they certainly do not make use of the new possibilities of expression favoured by parametric programmes, and by modellers of forms. The current vogue for rounded corners, curved surfaces, handles and protuberances of a quasi-biological origin, seems for the time being not to affect figurative research of architectonic origin. However, there are trends which cannot be concealed and the tendency to soften shapes, to cushion surfaces and break up geometric rhythms which are too rigidly defined, is increasingly evident and surely appears as the major trend of our time, the master movement which drives the figurative experimentation of these years.

Meanwhile, minimalism also appears to be taking on a new lease of life – now ever more curtailed and abstract, it has opened the way to a new expressiveness where formal research is combined with a refined selection of materials, textures and flavours. This new direction derives from the desire to make contemporary things stand out from the ordinary, to decontextualize them where possible, with a healthy and authentic touch of irony and disenchantment; it may, however, at times appear to be a banal and empty desire to poke fun and it can also be difficult to distinguish between what has a meaning and what is a hackneyed joke.

There is no lack of big names on the list of this year's chosen designers: Sottsass, Starck, Citterio, Cibic and many others who also remain meaningful references both in experimental projects and commercial products. Equally significant is the presence of Philips, Siemens and Sharp, whose design centres have developed a vast volume of work over the past few years, offering products on the market of extraordinary figurative quality and opening up to consumers products and environments updated in terms of both image and style.

In conclusion, I would like to address myself to those whose work has not been included: limited space has prevented it. I would, however, like to tell them not to worry and not to be offended – firstly because I am often wrong and secondly because what is culturally diffused is not always genuinely deserving of discriminating commercial success. I only hope that this selection adds to the quality on the list of all the *International Design Yearbooks* published up to now; I hope above all that this edition will also prove, as the others have for me, a favourite source of inspiration. Everything that can stimulate creativity deserves to be propagated.

[Translated from Italian by Carmona UK Ltd]

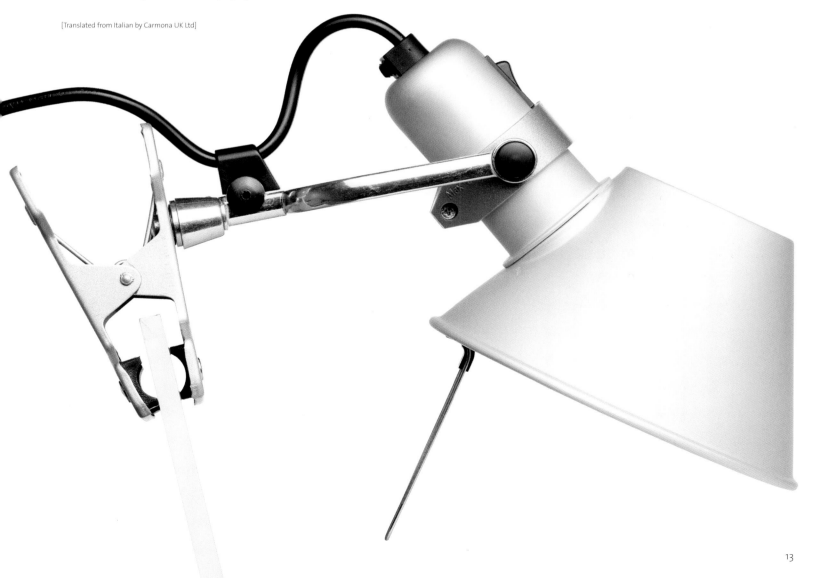

Michele De Lucchi
Lamp, Tolomeo Pinza
Artemide, Italy
1996

Furniture

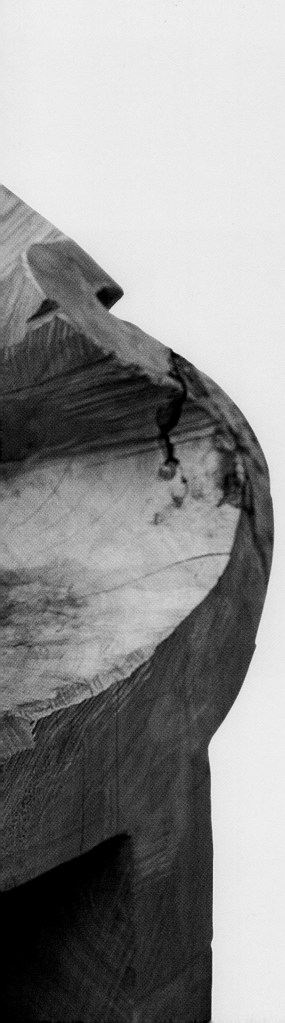

The Milan Furniture Fair played safe in 2000. The majority of designs on offer were still controlled by the dictate of profitability. There was generally more colour on show than the previous year but on closer examination these examples proved to be, on the whole, re-editions or pop culture reinterpretations. It was telling that post-show the media picked out Pesce's re-edition of the red monolithic 'Up' series for Baleri Italia; the 1970s nostalgia of Ross Lovegrove's 'Air One' stacking chair and stool in Biba colours; Philippe Starck's reworking of the Emeco aluminium chair; and the opening of the Kartell Museum, charting the company's history from its pioneering work in plastics to the present day, as the major events of the week.

As always, the off-site shows and the Salone Satellite offered greater variation than the main event. Here one saw collections organized by entrepreneurs willing to take risks. Although Cappellini played the theme music to 'A Clockwork Orange' at its show, bringing to mind the film's futuristic pop-culture connotations, the pieces exhibited by Barber Osgerby, the Bouroullec brothers, Mauro Mori, Jasper Morrison and Alfredo Haberli, amongst others, were fresh and innovative. Leon van Gerwen presented 'Hidden', a collection of home furnishings from a group of young individuals who were allowed to create without consideration of commercial pressure. Droog again reinvented how we view design, developing a series of objects to be played with and on which we could imprint our own personalities. Sputnik, created by Teruo Kurasaki of Idee, Japan, offered design to the internet. A collection of items by Michael Young, Marc Newson, Tim Power, Emmanuel Babled and others was shown which normally would only be available via the net. Totem, founded by David Shearer, financed the G7 show. Shearer's aim is to promote American design, especially by young, up-and-coming designers – he has sought them out, paired them with manufacturers and marketed their lines, or produced them himself through Totem's in-house production team. The show reflected the fusion of art and commerce which typifies contemporary American design. Michele De Lucchi's selection in this volume reflects his interest in the appearance of these new manufacturing and marketing initiatives.

De Lucchi was also concerned to highlight the proliferation of home-office furniture on the market. He has included items which he considers offer new aesthetic values and improve the quality of the environment in which we live and work, such as Defne Koz's 'Plano', N2's 'Ajax' for ClassiCon and Dante Donegani and Giovanni Lauda's designs for Radice. Young designers with experimental ideas and who have set-up their own companies or studios also feature. Pascal Tarabay's work is of particular interest with its combination of irony and function. The 'Frog' chair resembles a plastic garden chair but is in fact made from wood and the 'Beirut' bench challenges the formality of that archetype with the option to recline and linger a while. Lastly, craft appears in the choice. Here De Lucchi believes that the creative impulse is set free from restraints of market demand and the consequences of commercial success but it is from this freedom to experiment that he considers concepts for industrial prototypes could develop.

What of the future? Two important and not unrelated trends seem to be developing. Firstly, more than at any previous time the 'man-in-the-street' is talking about design – there is an unprecedented interest in home furnishing and we are deluged with lifestyle magazines. Design has become a popular movement. Every high street has a design shop and there is a greater demand for affordable and accessible contemporary design pieces. Many designers are appreciating this change: Authentics and Alessi lead the way with reasonably priced items, and Jasper Morrison, Mario Bellini, Matthew Hilton, Ron Arad and Philippe Starck have all produced furniture which can be bought without breaking the bank. Starck declares 'My job is to fight for utopia: popular designs and democratic designers. I want to give good design to as many people as possible. My most important fight is to take two zeros off the price.' Secondly, with the advance in digital technology, new forms and materials are being developed. Complicated and curved shapes can be created on CAD programs and a prototype can be printed out directly from a computerized milling machine. These rounded organic shapes lead to a demand for different, more pliable materials, such as foams, plastics, felts, cast aluminium and polypropylene, and with this a different emphasis on colour emerges. These materials in turn are easier and cheaper to produce – advantages which will filter down to the consumer. Eventually new technologies will succeed in increasing everbody's freedom to enjoy design at first hand.

Roderick Vos
Chair, Merak
Stainless steel, rattan
h. 110cm (43³/₈in) w. 81cm
(31⁷/₈in) l. 52cm (20½in)
Espaces et Lignes, Belgium
Limited batch production

Roderick Vos
Low chair, Sari
Stainless steel, rattan
h. 63.5cm (25in) w. 82cm
(32¼in) l. 69.5cm (27³/₈in)
Driade SpA, Italy
Limited batch production

Roderick Vos
Lounge chair, Kraton
Stainless steel, rattan
h. 76cm (30in) w. 81cm
(31⁷/₈in) l. 74cm (29⅛in)
Driade SpA, Italy
Limited batch production

Roderick Vos
Chair, Agung
Stainless steel, rattan
h. 75cm (29½in) w. 90.5cm
(35⅝in) l. 115cm (45¼in)
Driade SpA, Italy
Limited batch production

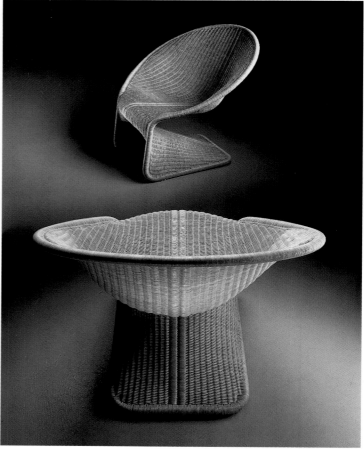

Karim Rashid
Chair, Wicker
Wicker
h. 101cm (40in) di. 56cm (22in)
Idee, Japan

Caroline Casey
Daybed, Zella
Seagrass, cane
h. 58cm (22⁷⁄₈in) w. 95cm (37³⁄₈in) l. 202cm (79¹⁄₂in)
Limited batch production

Ross Lovegrove
Chaise longue, Chaise Lounge
Loom membrane, aluminium, steel/inox
h. 25cm (9⁷⁄₈in) w. 80cm (31¹⁄₂in) l. 185cm (72⁷⁄₈in)
Loom, Germany

Ross Lovegrove
Table
Aluminium, glass, steel/inox
h. 72cm (28³⁄₈in) di. 110cm (43³⁄₈in)
Loom, Germany

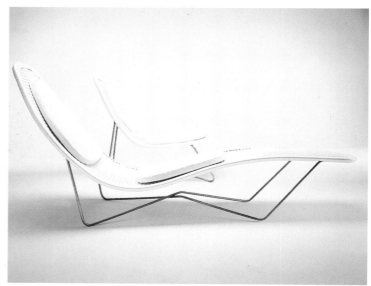

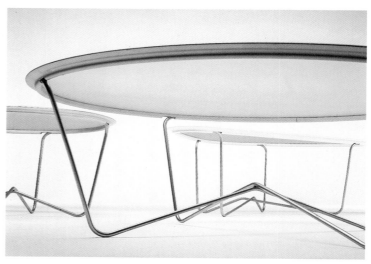

Godobert Reisenthel of Loom is the German importer of Lusty Lloyd Loom. Loom does, however, produce its own pieces as well, and the Ross Lovegrove collection is an example of a traditional manufacturer becoming involved with contemporary design. Reisenthel was impressed by the way Lovegrove mixed archaic, human shapes with a modern technical expression of form. The designs were created on computer and aluminium extrusions were added to the traditional paper and wire fabric.

Jane Dillon, Tom Grieves
Chair
Lloyd Loom fabric, plastic
h. 79.2cm (31in) w. 71cm (28in) d. 54cm (21¼in)
Lusty Lloyd Loom, UK
Prototype

Gitta Gschwendtner
Chair
Lloyd Loom fabric, metal
Lusty Lloyd Loom, UK

Forever associated in the British consciousness with the summer houses of a fading empire, Lloyd Loom furniture, traditionally made from woven twisted paper reinforced with steel wire stretched and attached to a rattan or beech frame, is being given a facelift. Following on from Nigel Coates's designs for Lloyd Loom of Spalding, Lusty Lloyd Loom has commissioned new designs from members of the Royal College of Art, London: Jane Dillon (Tutor of Design Products) and former student Gitta Gschwendtner. Building on the enormous technological advances the company originally made in order to work with the material, the new collection has also led to the development of innovative techniques. Jane Dillon's designs have made an important breakthrough, the fabric shaped by specially designed moulds which can be attached to a simple frame. These frames are constructed in a plastic manufactured to resemble wood, or metal, which has the advantage over rattan of being more pliable and waterproof.

Paola Navone
Three door partition, Black 90
Bamboo
h. 190cm (74⁷⁄₈in) l. 189cm
(74³⁄₈in)
Gervasoni SpA, Italy

Paola Navone
Seat, Black 02
Pulput, wood, teak
h. 110cm (43³⁄₈in) l. 175cm
(68⁷⁄₈in) d. 85cm (33¹⁄₂in)
Gervasoni SpA, Italy

Paola Navone
Screen, Black 99
Black bamboo, walnut
w. 150cm (59in) h. 180cm
(71in)
Gervasoni SpA, Italy

Mauro Mori
Chair, Round
Albizia wood
h. 40–60cm (15³⁄₄–25¹⁄₄in)
di. 60cm (23⁵⁄₈in)
One-off

Paola Navone
Seat, Malaka 02
Malacca, rawhide
h. 62cm (24³⁄₈in) w. 126cm
(49¹⁄₂in) d. 88cm (34⁵⁄₈in)
Gervasoni SpA, Italy

Natanel Gluska
Chair
Beech
One-off

To emphasize his belief in the importance of craft, De Lucchi has placed together designs which have a certain 'ethnic' feel either in the materials that have been used – wood, rattan etc. – or form.

Gluska's chairs are all unique. He attacks a virgin tree stump with a chain saw and in one piece fashions from it a hybrid of sculpture and functional item. De Lucchi states that crafts 'combine technology and human talent, knowledge and sensitivity, know-how and skill' (see p12). Here, inhibitions, whether creative or market-led, are shed allowing the designer real freedom of expression.

Cecilia Enevoldsen, Mark Burer
Stool
Maple
h. 45cm (17³/₈in) w. 30cm (11⁷/₈in)
d. 30cm (11⁷/₈in)
Prototype

Louise Campbell
Chair, Ho'nesty
Ash
d. 65cm (25¹/₂in) di. 130cm (51¹/₈in)
One-off

Komplot Design
Deckchair, Clinker
Ash
h. 100cm (39³/₈in) w. 80cm (31¹/₂in) l. 140cm (55in)
Komplot Design, Denmark
One-off

Hans Sandgren Jakobsen
Stool, The Rockable
Ash
h. 48.4cm (19in) l. 76cm (30in) d. 38cm (15in)
Andre Skriver, Denmark
One-off

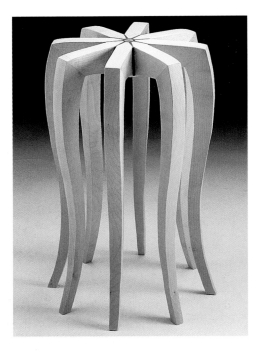

The synergy of tradition and modernity is beautifully illustrated in the Kunstindustrimuseet exhibition, 'Walk the Plank'. Twenty designers were asked to work in partnership with 20 cabinet makers to create a series of furniture out of a wooden plank. Following the show, the pieces were auctioned, with the proceeds being used to subsidize future prototypes. Michele De Lucchi was particularly interested in the project, seeing it as genuinely interested in design as culture while still supporting the development of the furniture industry.

Hans Sandgren Jakobsen
Stool, The Unrockable
Ash
h. 48.4cm (19in) l. 76cm (30in) d. 38cm (15in)
Andre Skriver, Denmark
One-off

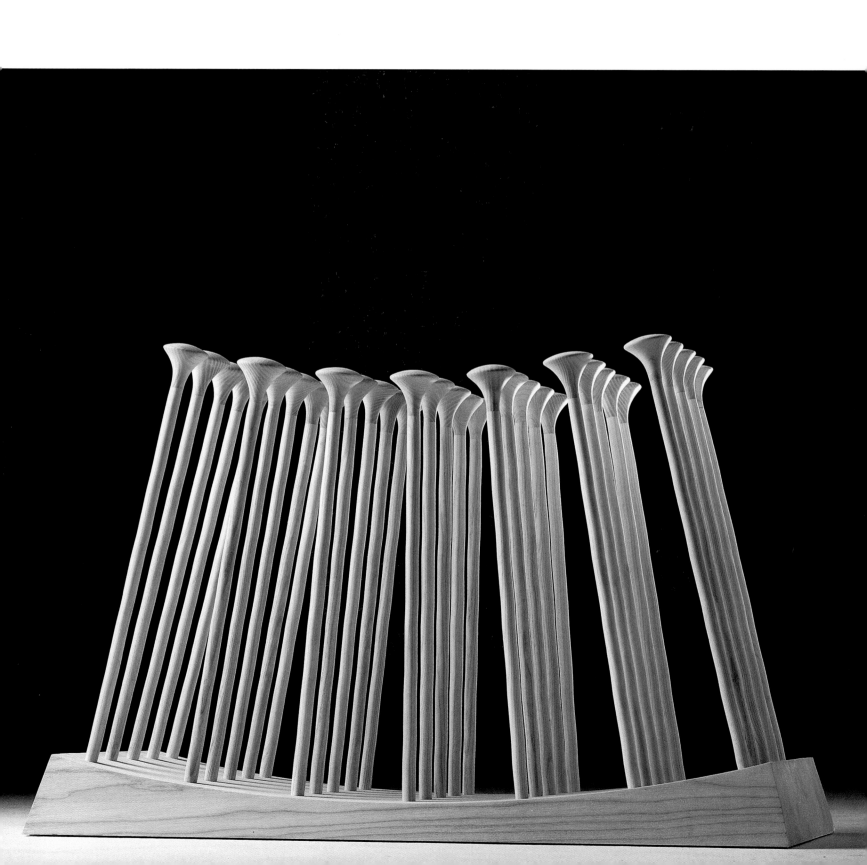

Henrik Schulz
Chair, Rubber chair
Rubber
h. 55cm (21⅝in) w. 73cm (28¾in) l. 62cm (24½in)
-ing, Denmark
Prototype

Henrik Schulz
Chair, Modern rocker
Steel tubes, leather
h. 58cm (22⅞in) w. 65cm (25⅝in) l. 82cm (32⅛in)
-ing, Denmark
Prototype

Todd Bracher
Café Environ, Open Privacy
Ash ply veneer
h. 180cm (71in) w. 180cm (71in) l. 180cm (71in)
-ing, Denmark

Hanspeter Steiger
Chair, Loi
Plywood
h. 78cm (30¾in) w. 41cm (16⅛in) l. 45cm (17¾in)
-ing, Denmark
Prototype

Dögg Gudmundsdóttir
Chair/chaise longue, Wing
Plywood
h. 125cm (49⅛in) w. 47cm (18½in) l. 59cm (23⅛in)
-ing, Denmark
Prototype

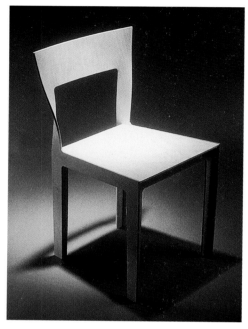

-ing is a young international collaborative which has come together to create a series of bold and stimulating yet functional pieces. Its members, Dögg Gudmundsdóttir (Iceland); Todd Bracher (USA); Hanspeter Steiger (Switzerland); and Henrik Schulz (Sweden) come from varying backgrounds, ranging from architecture and carpentry to different fields of design – industrial, furniture and graphic. This wealth of influence informed its first collection, shown in the Salone Satellite in 2000. 'Open Privacy's' architectonic form has created a room within a room while 'Wing' and 'Loi' show -ing's concern to liberate the full potential of a material – plywood is moulded into a 'straight' chair with an unusual 'twist' and a chair/chaise longue fits ergonomically to the body.

Ely Rozenberg
Chaise longue, Moby
Harmonious steel, zipper
h. 63cm (24⁷⁄₈in) w. 60cm (23⁵⁄₈in) l. 160cm (63in)
oz, Italy
Limited batch production

Ely Rozenberg
Armchair, Poltronalampo
Harmonious steel, zipper
h. 90cm (35½in) w. 60cm (23⁵⁄₈in) l. 65cm (25⁵⁄₈in)
oz, Italy
Limited batch production

In his introduction to *The International Design Yearbook 1999*, Jasper Morrison expressed what he believed made an experimental design work: 'the best of them contain a conceptual element which makes use of forms borrowed and adapted from different applications and a subtle play with materials which, combined with the right emphasis, bring new and interesting results'. This could have been written to accompany Ely Rozenberg's zipper chairs. Rozenberg has lazer-cut 0.6mm steel – used in the production of springs – into a template; strong adhesive has then been applied to the edges of the pattern, the zipper added and the whole neatly assembled in one easy movement. Due to this new technique, the pieces can be opened up, disassembled and stored easily. Steel, normally rigid and almost always soldered or riveted, is processed here as if it were a tailored fabric. This imaginative fusion of an industrial substance with an invention handed down from the last century challenges current conventions of material combination.

Riccardo Blumer
Chair, Laleggera
Wood, polyurethane, nylon
h. 79cm (31⅛in) w. 44cm (17⅜in) d. 53cm (20⅞in)
Alias srl, Italy

Riccardo Blumer
Laleggera stacked range
Tables, Ilvolo
Maple/Ash, polyurethane, nylon
h. 73cm (28½in) l. 90–220cm (35⅜–86⅝in) w. 90cm (35⅜in)
Alias srl, Italy

Riccardo Blumer
Stool, Laleggera stool
Wood, polyurethane, nylon
h. 44cm (17⅜in) w. 36cm (14⅛in) d. 35cm (13¾in)
Alias srl, Italy

Hannes Wettstein
Chair, Alfa
Aluminium
h. 80cm (31½in) l. 49cm (19⅜in) d. 50cm (19⅝in)
Molteni & C SpA, Italy

Pascal Mourgue
Chair, Smala
Steel, aluminium
h. 82cm (32¼in) w. 65cm (25½in) d. 62cm (24⅜in)
Ligne Roset, France

Konstantin Grcic
Chair, Allievo
Beech, plywood
h. 97cm (38⅛in) l. 54cm (21¼in) d. 87cm (34¼in)
Montina International srl, Italy

Konstantin Grcic
Chair, Scolaro
Beech, plywood
h. 75cm (29½in) l. 54cm (21¼in) d. 87cm (34¼in)
Montina International srl, Italy

Michele De Lucchi wanted to group together some of the minimalist designs he selected. Although not all the minimalist designs could be grouped together, the point can be made here: minimalism does not go away. It has been a significant development in the twentieth century and is growing stronger in the twenty-first. By honing down the decorative element in a design a greater importance can be placed on material and form, offering the designer the opportunity to express himself in the purest terms.

Perry King, Santiago Miranda
Stackable chair, Lisa
Thermoplastic technopolymer, aluminium, steel, rubber
h. 85cm (33½in) w. 44cm (17³⁄₈in) d.44cm (17³⁄₈in)
Baleri Italia SpA, Italy

Peter Wheeler, Mary Little
Outdoor stool, Lulu
Terracotta
h. 22cm (8⅝in) di. 60cm (23⅝in)
Bius, UK
Prototype

Gabriela Nahlikova, Leona Matejkova
Chair, Airchair
Inflatable cushions, stainless steel
h. 85cm (33½in) w. 38cm (15in) d. 39cm (15³⁄₈in)
Prototype

Shin Azumi, Tomoko Azumi
High stool, LEM
Steel, plywood
h. 65–75cm (25⅝–29½in) w. 39cm (15⅜in) d. 42cm (16½in)
Lapalma, Italy

Gabriela Nahlikova, Leona Matejkova
Chair, Sesle
Beech, silk
h. 82cm (32in) w. 42cm (16in) d. 39cm (15in)
Prototype

After studying at the Royal College of Art, London, Shin and Tomoko Azumi founded their studio in 1995. They were both born in Japan in the mid-1960s and studied product and environmental design respectively. Their creations are timeless and practical. Early pieces illustrated their belief that design should be accessible to all and solve problems of everyday living. The transforming objects of the late 1990s refused to limit definition by combining two or three uses in one, creating furniture for use in limited space while the 'Wire Frame' series used supermarket-trolley technology to produce an inexpensive chair and bench. Later mass-produced objects, such as the stools above, all share a simplicity of and attention to detail. 'We want to concentrate on the functional aspects of design, not its decorative features,' said Tomoko.

Emilio Ambasz
Stacking chair, Vox
Steel frame, aluminium
h. 79cm (31⅛in) w. 53cm (20⅞in) d. 47cm (18½in)
Vitra, Switzerland

Vardit Laor
Chair Cabinet
Plywood, glass
h. 285cm (112¼in) w. 80cm (31½in) d. 40cm (15⅝in)
Limited batch production

François Azambourg
Chair, Chauffeuse et Pouf
Birch plywood, natural foam rubber
h. 71cm (28in) w. 70cm (27½in) l. 97cm (31¼in)
VIA, Paris
Prototype

One of the more exciting shows in Milan 2000 was VIA's 'Design France – Generation 2001', the 20th anniversary exhibition of VIA (Valorization of Innovation in Home Furnishings). VIA was created in 1979 by the Committee for the Development of French Furniture Industry (CODIFA) and the Ministry of Industrial Production, and today its importance to French furniture manufacture is indisputable. Its functions are manifold: to gauge trends based on changes in lifestyle, collaborate with the various art schools in France, promote French design, act as a reference body of creative talent working within France (Agora des Createurs) and advise manufacturers in the field on matters of strategy and product development. Most importantly for the aspiring younger generation of designer, its 'Appels Permanents' commission looks at over 1,000 portfolios each year and finances prototypes for the most outstanding projects. The commission also awards a research grant, the 'Carte Blanche', to individuals who it considers show both originality and a mature, creative approach. Few other countries seem to have an equivalent organization dedicated to promoting the innovative character of their home furnishing industry to that which VIA is offering France.

Ross Lovegrove
Chair, Air One
Polypropylene foam
h. 53cm (20⅞in) d. 115cm (45in)
Edra, Italy

Fabricated from a polypropylene foam normally used in packing, Ross Lovegrove's 'Air One' stacking chair and stool (not shown) in disco blue and silver recapture the zeitgeist of the 1970s. They are extremely light and can be transported and stacked easily.

Danny Lane
Table, Laughing Water
Glass, stainless steel
h. 72cm (28½in) w. 170cm (67in) l.193cm (76cm)
One-off

One Foot Taller
Chair, Chasm
Polyetheline
h. 75cm (29½in) w. 70cm (27½in) d. 75cm (29½in)
Nicehouse, UK
Limited batch production

One Foot Taller
Chair, Ravine
Polyetheline, stainless steel
h. 80cm (31½in) w. 50cm (19½in) d. 55cm (21⅝in)
Nicehouse, UK
Limited batch production

Ron Arad
Sofa, Victoria and Albert Collection
Polyester resin, steel, polyurethane foam
h. 85–145cm (33½–57in) w. 180–290cm (70⅞–114⅛in)
Moroso SpA, Italy

Ron Arad
Armchair, Victoria and Albert Collection
Polyester resin, steel, polyurethane foam
h. 75cm (29½in) w. 74cm (29⅛in)
Moroso SpA, Italy

Ron Arad's Victoria and Albert sofa for Moroso is named after the London museum where an exhibition of Arad's work ran from June to October 2000. It is formed from a single band of tempered steel which is clad in foam and upholstered in a washable but irremovable cover in primary colours. It can be made in any size although the bigger the better. Javier Mariscal said just after he had seen the sofa at the Moroso show in Milan that its size, luxuriance and sensuous curves gave it an indecent air – it is a love couch to be played on and enjoyed. This is a rather pertinent comic comment, given that Professor David Starkey (visiting fellow at Fitzwilliam College, Cambridge) revealed recently, while making a documentary on the life of Queen Victoria, that, contrary to popular belief, the queen had enjoyed a boisterous, fun-loving attitude to sex.

Philippe Starck
Chair, Hudson
Highly polished aluminium
h. 84cm (33in) w. 42cm (16⁵⁄₈in) d. 46cm (18¹⁄₄in)
Emeco, USA

Pascal Tarabay
Chair, Frog
Wood, aluminium, plastic
h. 84cm (33in) w. 50cm (19⁵⁄₈in) l. 50cm (19⁵⁄₈in)
Limited edition

Mario Bellini
Chair, Arco
Fibreglass, polymer
h. 85cm (33¹⁄₂in) w. 63.5cm (25in) d. 54.5cm (21¹⁄₂in)
Heller Inc., USA

Karim Rashid
Chair, Oh Chair
Polypropylene, steel
h. 86.3cm (34in) w. 61cm (24in) d. 56cm (22in)
Umbra, Canada

Philippe Starck had always been impressed with the Emeco aluminium chair and – unusually for Starck – approached the company asking if he could be responsible for its remodelling. He says, 'the Emeco chair is a timeless design but you can see it comes from the 40s or 50s. I wanted to keep the heritage of the chair, but turn it into an almost future classic'. He kept the main silhouette of the original while making it stackable, lighter and better adapted to today's needs.

Karim Rashid is undoubtedly the best-known contemporary designer working in the USA today. His expertise crosses the boundaries of lighting, tableware and furniture design and is admired internationally. His use of fluid, sensual curves and command of material – be it Murano glass or latest recyclable polypropylene – has resulted in an eclectic portfolio of work. The 'Oh Chair' is an ergonomic and kinaesthetic multipurpose stacking chair. The concept behind its creation was to produce an elegant inexpensive design. The employment of flexible polypropylene means that 'the man in the street' can afford to buy a plastic chair which is actually comfortable to sit on.

Gunilla Allard
Chair, Cosmos
Steel, wood
h. 80cm (31½in) w. 36cm (14⅛in) d. 50cm (19⅝in)
Lammhults Mobel AB, Sweden

Kasper Salto
Stacking chair, Blade
Birch
h. 79cm (31⅛in) w. 49cm (19¼in) l. 50cm (19⅝in)
Botium, Denmark

Alberto Meda
Office chair, Meda 2
Plastic, net weave
h. 96–101cm (37¾–39¾in) w. 67cm (26⅜in) d. 67cm (26⅜in)
Vitra, Switzerland

Emilio Ambasz
Stacking chair, Vox
Steel frame, aluminium
h. 79cm (31⅛in) w. 53cm (20⅞in) d. 47cm (18½in)
Vitra, Switzerland

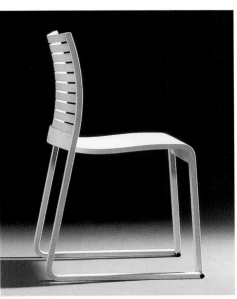

Richard Hutten
Chair, One of a Kind
Alucobond, aluminium
h. 45–82cm (17³/₄–32¹/₄in) w. 40cm (15³/₄in)
sdb Industries, The Netherlands

Jörg Boner
Cupboard, Hoover
Aluminium, fabric
h. 199cm (78³/₈in) w. 67cm (26¹/₄in) d. 57.5cm (22⁵/₈in)
sdb Industries, The Netherlands

Richard Hutten
Lounge chair, One of a Kind
Alucobond, aluminium
h. 69cm (27¹/₈ in) w. 67cm (27in) d. 50cm (19⁵/₈in)
sdb Industries, The Netherlands

Group Kombinat
Book cupboard, High in the Sky
MDF, veneer, aluminium
h. 42cm (16¹/₂in) w. 40cm (15³/₄in) d. 40cm (15³/₄in)
sdb Industries, The Netherlands

Group Kombinat
Lounge Chair, Missy
Aluminium, polyurethane
h. 104cm (41in) w. 60cm (23⁵/₈in) d. 60cm (23⁵/₈in)
sdb Industries, The Netherlands

Geert Koster
Storage Unit, Reflex
Multiplex, plexi, aluminium
h. 160cm (63in) w. 215cm (84⁵/₈in) d. 70cm (27¹/₂in)
sdb Industries, The Netherlands

Group Kombinat
Seat with light, Zebra
MDF, veneer, fabric
h. 48cm (18⁷/₈in) w. 140cm (55¹/₈in) d. 49cm (19¹/₄in)
sdb Industries, The Netherlands

Ron Arad
Table, No Waste
Metawell
h. 75cm (29½in) w. 220cm (86⅝in) d.130cm (51⅛in)
sdb Industries, The Netherlands

Richard Hutten
Cabinet, Wheels
MDF, steel, aluminium
h. 150cm (59in) w. 100cm (39⅜in) d. 70cm (27½in)
sdb Industries, The Netherlands

Geert Koster
Side table, 3 Headed Table
Alucobond, steel
h. 60cm (23⅝in) di. 35cm (13¾in)
sdb Industries, The Netherlands

Leon van Gerwen is an enlightened entrepreneur – genuinely enthusiastic about promoting contemporary design. He joined sdb as marketing manager in 1996 when the company controlled the middle segment of the display cabinet market but when its profitability was declining. It was at this time that the seeds of 'Hidden' were planted. He came up with the idea of a collection of original products by a group of young designers who shared the same taste, ideas and visions. So far nine individuals/studios have been commissioned: N2, Group Kombinat, Ron Arad, Richard Hutten, Geert Koster, Dumoffice, El Ultimo Grito, Atelier Oi and Christophe Pillet.

Why 'Hidden'? Van Gerwen is concerned that the collection should be what his designers believe the market wants. Too often manufacturers 'hide' designs away, thinking them to be uncommercial. 'Innovation is important and you must not immediately shout that something cannot be made or that the market is not yet ripe for it. I want to show the public the pure design, irrespective of the costs,' he says. His role is to produce the technology to make the collection work economically without losing sight of the creation itself and to find the right distribution channels.

I asked van Gerwen about his vision. I could understand, and admire the concept of 'Hidden' but couldn't quite see how Arad's 'Reinventing the Wheel' (not shown) fitted into the pattern, having been such a success in Milan in 1996. Van Gerwen replied that the first pieces of this design were made in steel by Ron himself; now, through his company, it was available in a smaller version and in aluminium at a fraction of the price, making it more 'visible'. 'Hidden' gives hidden talent a chance in more ways than one.

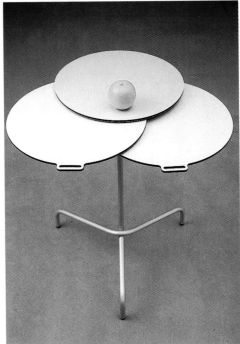

Giancarlo Piretti
Folding chair with oscillating back-rest, Torsion
Aluminium, polypropylene
h. 81cm (31⅞in) w. 47cm (18½in) d. 58.5cm (23in)
Prototype

Alberto Meda
Seating system, Floating Frame
Aluminium, polyester net
Various sizes
Alias srl, Italy

Verner Panton
Chair, Pantostack
Diecast aluminium, polypropylene
h. 84cm (33in) w. 65cm (25½in) d. 55cm (21⅝in)
VS Vereinigte Spezial Mobelfabriken GmbH & Co, Germany

The Pantostack was Verner Panton's last furniture design. Although he died in September 1998, production tests meant that manufacture did not take place until late 1999.

Erwan Bouroullec
Armchair, Spring
Steel, fibreglass, polyurethane, rubber
h. 36–68cm (14⅛–26¾in) w. 72cm (28⅜in) d. 80cm (31½in)
Cappellini SpA, Italy

Mauro Mori
Service table, M2544
Marble
h. 36cm (14⅛in) di. 33cm (13in)
Cappellini SpA, Italy

Barber Osgerby
Pouf or table, Hula
Teak heartwood
h. 34cm (13⅜in) w. 40cm (15¾in) l. 75cm (29½in)
Cappellini SpA, Italy

Mauro Mori
Service table, M2546
Rose Albitia wood
h. 40cm (15¾in) w. 61cm (24in) l. 38cm (15in)
Cappellini SpA, Italy

Alfredo Häberli
Pouf, PO/0028
Polyurethane foam
h. 42cm (16½in) di. 43cm (17in)
Cappellini SpA, Italy

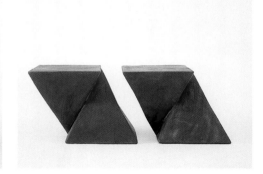

Cappellini SpA continues to stun us with its new and exciting designs which are collected from designers around the world. A team of five sieves through hundreds of products presented to the company every year. All pieces selected fit into a corporate identity and each is directly approved by Giulio Cappellini himself. He states that a design must have an innovative form and/or use of material to enter into his collection. 'Each product must do something better than the one before,' he says.

I asked him whether the use of 'big names' and relative newcomers was calculated, and if guaranteed successes from Lissoni, Morrison and so on left him free to promote more experimental designs. He replied: 'Products do not have priority for me. They have the same value. I can sell items from both famous designers and young individuals. Each prototype takes years of development before it reaches a collection and then often many products ripen very slowly on the market but this doesn't matter, since at Cappellini we like to work more on long-sellers than best-sellers.'

He recognizes that there are very few companies willing to work with more initiatory designs and that this is to blame for the lack of choice in the market today. He says, 'the designers of the future will be the ones that not only project good products but also create a good atmosphere. They will create "human" projects that will let the user dream. I feel that some of the designers cooperating with Cappellini are working in this direction.'

Cappellini is undoubtedly an entrepreneur. The company exports to more than 50 countries, employs hundreds of people and its turnover increases yearly. However, Cappellini's poetic eye and cognizance of what is fresh, plus his willingness to give new designs a chance, guarantees his collections are not only commercially successful but also varied and beautiful.

Jasper Morrison
Stackable chair, Tate
Beech plywood, oak, polypropylene
h. 45-81cm (17³/₄–31⁷/₈in) w. 53cm (20⁷/₈in) d. 47.5cm (18³/₄in)
Cappellini SpA, Italy

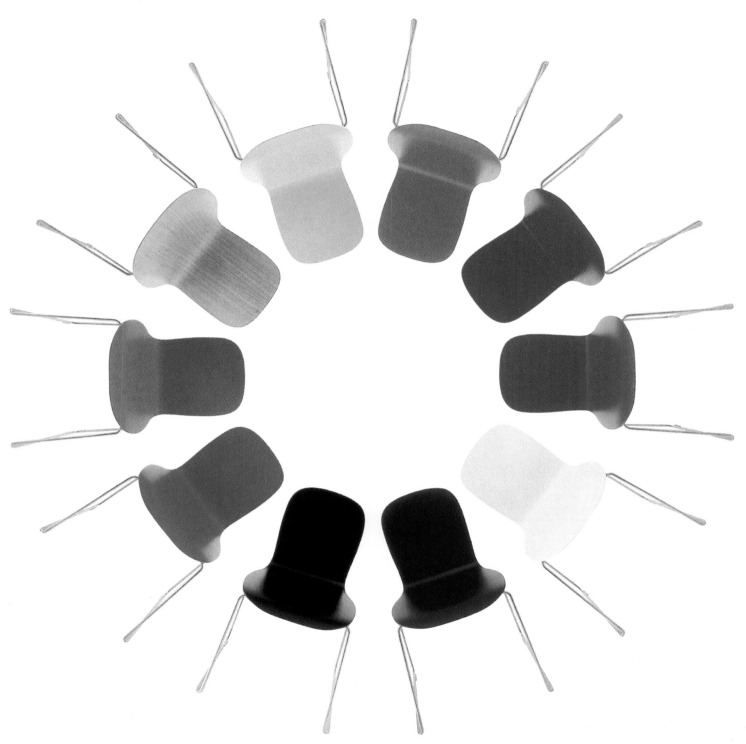

Furniture

Alfredo Häberli
Extendible table, Easy Long
Nickel, oak, beech
h. 73cm (28¾in) w. 130cm (51⅛in) l. 190–250cm
(74⅞–98½in)
Cappellini SpA, Italy

Piero Lissoni
Cabinet, Uni
Metal, lacquer, macroter
h. 64cm (25⅛in) l. 270cm (106¼in) w. 61.2cm (24⅛in)
Cappellini SpA, Italy

Claudio Silvestrin
Table and bench, Millennium Hope
Walnut heartwood
Table: h. 72cm (28⅜in) l. 315cm (124in) d. 80cm (31½in)
Bench: h. 45cm (17¾in) l. 157.5cm (62in) d. 40cm (15¾in)
Cappellini SpA, Italy

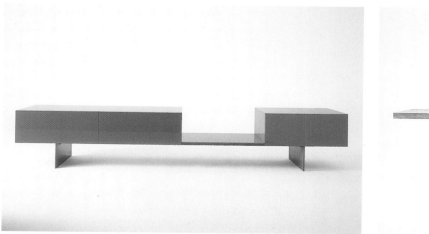

Erwan Bouroullec
Bed compartment, Lit Clos
Steel, birch plywood, aluminium, metacrylate,
cellulose, fabric
h. 75–324cm (29½–127½in) w. 240cm (94½in)
d. 200cm (78¾in)
Cappellini SpA, Italy

Carlo Colombo
Storage system, Archi
Stainless steel, glass, wood
Various sizes
Cappellini SpA, Italy

Kasper Salto
Daybed for a child, Leaf
Birch sheet
h. 15cm (6in) w. 35cm (13¾in) l. 85cm (33½in)
Prototype

Pascal Tarabay
Bench, Beirut
Metal, wood or recycled plastic
h. 75cm (29½in) w. 180cm (70⅞in) l. 120cm (47¼in)
Limited edition

Pascal Tarabay
Chaise longue, Lazy lounge
Metal, wood, EVA foam
h. 50cm (19⅝in) w. 65cm (25⅝in) l. 180cm (70⅞in)
Limited edition

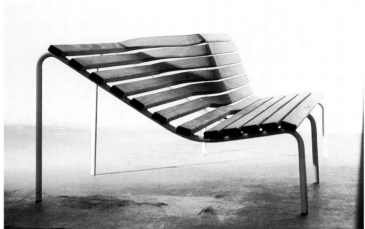
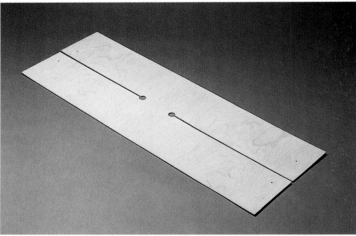

Kasper Salto started his daybed-for-a-child design on paper. He worked on the assumption that what can be made with paper can also be made with plywood, so cut out a rectangle and made two slits in it, one in either end, creating two flaps. These ends were then pressed together and riveted in place, one forming the backrest and the other the leg support for the child.

Antonio Citterio
Sofa, Freetime
Chrome
h. 62cm (24½in) l. 174.5cm (68¾in)/259.5cm (102¼in)
d. 107cm (42⅛in)
B&B Italia, Italy

Piero Lissoni
Sofa, Metro 2
Wood, metal, white/orange fabric
h. 71cm (28in) l. 160–210cm (63–82½in) d. 100cm (39⅜in)
Living Divani, Italy

Werner Aisslinger
Storage system, Cell System
Cut crystal, aluminium
h. 80cm (31½in) w. 36cm (14⅛in) d. 50cm (19⅝in)
Zeritalia, Italy

Cini Boeri
Sofa, Meter
Polyurethane, foam, feathers, metal
h. 76cm (30in) w. 220/260cm (86⅝/102⅜in) d. 97cm (31⅛in)
Molteni SpA, Italy

Emaf Progetti
Sofa, Alfa
Steel, elastic, polyurethane, nylon, goose down
h. 65cm (25⅝in) l. 230cm (90½in) d. 92cm (36¼in)
Zanotta, Italy

Alfredo Häberli
Cabinet, Florence
MDF, aluminium alloy, steel
Various sizes
Zanotta SpA, Italy

Lodovico Acerbis
Cabinet, The Jolly Units
Wenge, cherry or lacquer finish
h. 161cm (63³/₈in) l. 120cm (47¹/₄in) d. 41cm (16¹/₈in)
Acerbis International SpA, Italy

Lodovico Acerbis
Shelves, The Shelvings
Wenge, cherry or lacquer finish
h. 220cm (86⁵/₈in) w. 80–120cm (31½–47¹/₄in) d. 35cm (13⁷/₈in)
Acerbis International SpA, Italy

Lodovico Acerbis
Shelves, Wall Shelves
Wenge, cherry or lacquer finish
h. 6cm (2³/₈in) d. 30cm (11⁷/₈in)
Acerbis International SpA, Italy

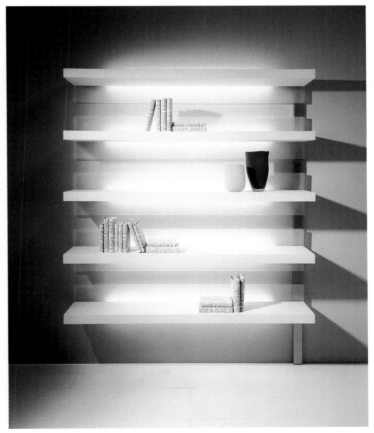

Chiara Cantono
Bidimensional furniture, Mat
Wood, moquette
h. 350cm (137³⁄₄in) w. 360cm (141³⁄₄in) d. 2cm (³⁄₄in)
Prototype

The young Italian furniture designer Chiara Cantono has achieved a new living concept. When not needed, the volume of the furniture is reduced to a plywood 'carpet' which can then be manipulated in a few seconds, as required, to form a series of living spaces; the elements are secured by a self-blocking mechanism. Reminiscent in its construction of Pietro Arosio's Mirandolina chair for Zanotta, which needed an industrial process to bend and stamp the pre-cut flat aluminium into a chair at one stroke, the beauty of this design is that it can revert simply to its flat-pack state.

Shigeru Uchida
Cabinet, Horizontal
Oak, aluminium
h. 91cm (35⅞in) w. 90cm (35½in) l. 18cm (7in)
Build, Japan
Prototype

Rodolfo Dordoni
Bed, Favignana
Maplewood
h. 158cm (62in) l. 200cm (78¾in) w. 171/181/191cm (67³/₈/71¼/75in)
Flou SpA, Italy

Patrizia Scarzella
Bed, Meridiana Testata Alta
Aluminium, wood
l. 200cm (78¾in)/210 cm (82⅝in) w. 160cm (63in)/170cm (67in)/180cm (70⅞in)
Flou SpA, Italy

Hans Sandgren Jakobsen
Bed, Grandlit
Beech, steel
h. 78cm (30¾in) w. 220cm (86⅝in) l. 200cm (78¾in)
Fredericia Furniture A/S, Denmark

Hans Sandgren Jakobsen's 'Grandlit' is a new concept in bed design. The table and two back supports can be placed anywhere around the perimeter, allowing for a variety of uses.

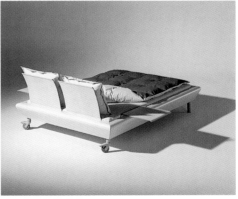

Boum Design
Lounge Chair, Air Lounge
Vinyl, elastic
5 cushions of h. 20.3cm (8in) w. 61cm (24in) l. 61cm (24in)
Boum Design, USA

Boum Design
Screen, Air Wall
Plexiglass, vinyl, elastic
9 cushions: h. 203cm (80in) l. 182.8cm (72in) d. 25.4cm (10in)
Boum Design, USA

Boum Design
Cushion, Air cushion
Vinyl
h. 20.3cm (8in) w. 61cm (24in) l. 61cm (24in)
Boum Design, USA

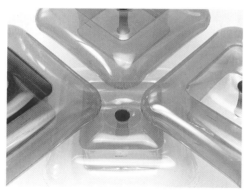

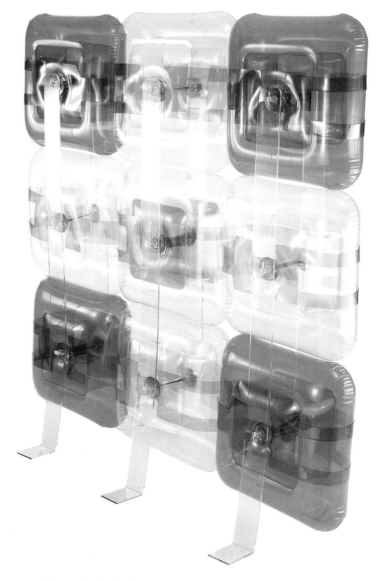

'The Americans are coming' was the cry in the streets during the Milan Furniture Fair in 2000. The G7 off-site show, where Pierre Bouguennec's (Boum Design) inflatable pieces formed an impressive centrepiece, was talked about with great excitement. The other members of the G7 group were Once (Marre Moerel, Harry Paul van Iersel and Camilla Vega), CCD (Christopher Deam), Prototype and Production (Chris Bundy and Ross Menuez), Worx (Michael Solis), Comma (David Khouri and Robert Guzman) and Dinersan Inc. (Nick Dine). Together their designs offer a new and rather light-hearted approach to design. Tactile and accessible, yet highly functional, they represent the beginning of a new American movement. Paola Antonelli of MoMA has said, 'A new generation of American designers has recently come to the forefront and will soon move the balance of creativity towards this side of the world ... it is a great moment for American design'.

Similarly, the Urburbia show in New York (see p61), running at the time of the International Contemporary Furniture Fair in New York 2000, highlighted the work of a group of young Americans (overlapping with some of the G7 show exhibitors). Nick Dine, Jeffrey Bernett, Richard Shemtov, Michael Solis, David Khouri and Harry Allen designed a range of furniture for city living. Shemtov – whose company, Dune, arranged the project – asked each to create pieces to 'incorporate size and utility with a sleek modern aesthetic'. The result, 'Urburbia', offers a cohesive alternative to the mix and match furniture of the average New York City-dweller.

Boum Design
Air Chair
Plexiglass, elastic
2 cushions: h. 61cm (24in) w. 81cm (32in) l. 81cm (32in)
Boum Design, USA

Jacob Timpe
Table, Tischbocktisch
Untreated ash, rubber
h. 73cm (28⁷⁄₈in) l. 190cm (74⁷⁄₈in) d. 85cm (33½in)
Moormann Möbel Produktions, Germany

Pascal Mourgue
Corner table, Smala
Oak, aluminium
h. 46cm (18in) l. 100cm (39³⁄₈in) w. 100cm (39³⁄₈in)
Ligne Roset, France

Piero Lissoni
Table, M.P
Lacquer/aluminium, wenge/beech
h. 24cm (9½in) w. 90cm (35½in) l. 130cm (51¹⁄₈in)
Artelano, France

Paolo Ulian
Coffee Table, Bench 2000
Plywood, aluminium
h. 30cm (11⁷⁄₈in) w. 75cm (29½in) d. 150cm (59cm)
Prototype

2000 saw the creation of a new prize – the Design Report Award promoted by the German magazine of the same name. The judges were Matali Crasset, Konstantin Grcic, Ross Lovegrove and Nasir Kassamali who, from the hundreds of designs offered by the Salone Satellite, picked Paolo Ulian's 'Bench 2000'. This coffee table is constructed from flexible plywood, and the top can be raised in one smooth action to form a bench with back support and integral storage space.

Jeffrey Bernett
Chair, Beta
Birch plywood, plastic laminate, ultrasuede
w. 49cm (19¼in) d. 51.5cm (20¼in) h. 82.5cm (32½in)
Dune, USA

Michael Solis
Cabinet, Four Forty
Dune, USA

View of the Urburbia show, New York
Michael Solis
Side Tables, Fuse+, Fuse-
Walnut veneer, matt lacquered MDF, aluminium
w. 41cm (16in) d. 41cm (16in) h. 60cm (23⅝in)
Dune, USA

Harry Allen
Bed, La La Salama
Walnut veneer, matt lacquered MDF, satin chrome
w. 223.5cm (88in) d. 51.5cm (20¼in) h. 228.5cm (90in)
Dune, USA

La La Salama means 'peaceful sleep' in Swahili. Harry Allen's Murphy bed not only looks good but is comfortable too. The prop of so many black-and-white comedies (where a scene of infidelity has to be hastily transformed for an unexpected guest by throwing everything, usually including discarded stockings, on to the bed and concealing them in the wall), this piece of Americana has been given a facelift. The walnut unit contains a pull-down queen-size bed and compartments hidden on one side by doors and on the other by a range of storage units, shelves and a mirror. A bedside table can also be lowered at night and pushed back into the unit during the day. As the bed is secured by legs to the floor and ceiling rather than a wall, it is ideal for use as a room divider.

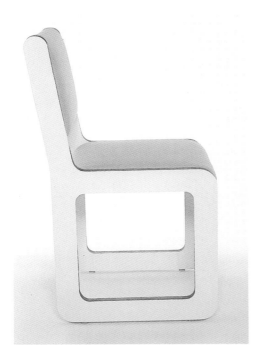

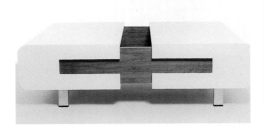

Antonio Citterio
Workstation, Vademecum
Steel, aluminium, MDF
h. 180cm (70⅞in) w. 120cm (47¼in) d. 105cm (41⅜in)
Vitra, Switzerland

Peter Maly
Office system, Modul 5000
Steel, lacquer, glass, maple, beech, walnut
Various sizes
Mauser Office, Germany

The home-office is a concept close to Michele De Lucchi's heart. His selection contains many examples of furniture which although designed to be worked at have also been modified visually and in some cases functionally to fit into the domestic environment. N2's and Defne Koz's desks re-interpret the traditional bureau (p64) as does Power's Go-Car (p66). By reducing the size of the work-station to the minimum and constructing it in brightly coloured glass, Power has created a new aesthetic in home/office typology. Radice's designs, on the other hand, neatly pack away the office when the work day is over (pp68–9). De Lucchi considers office furniture to be often too complicated whereas home/office products create an environment. He is presently working on the interior design of an office space where he intends to use only furniture which has been designed for the domestic workplace. Although he will employ matching pieces, he envisages a day when individual pieces will reflect a part of each worker's own character.

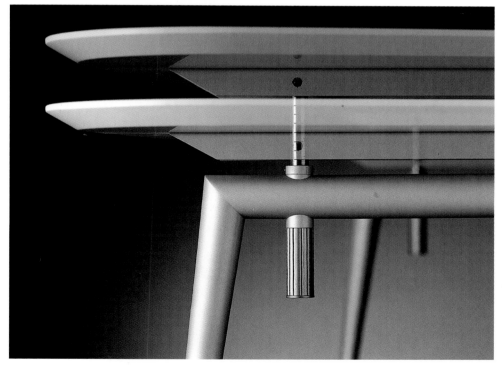

Christian Deuber, Jörg Boner (N2)
Table, Ajax
Wood, steel
h. 77cm (30¼in) w. 110cm (43¼in) d. 93cm (36½in)
ClassiCon, Germany

Defne Koz
Desk, Plano
Stainless steel, plywood
h. 92cm (36in) w. 105cm (41³⁄₈in) d. 90cm (35½in)
Mobileffe, Italy

Werner Aisslinger
Table, x-tisch
Wood, steel, aluminium
h. 70–78cm (27½–30⁵⁄₈in) w. 90cm (35½in) l. 220cm (86⁵⁄₈in)
Böwer GmbH, Germany

Werner Aisslinger's x-tisch is a multifunctional table which can be used both in the office or home. It is both foldable and height-adjustable, its stability being safeguarded by the x-tisch legs which move as the height of the table is adjusted.

Benjamin Thut
Folding table, Scissor table, Alu 1
Synthetic resin, refined steel
w. 80cm (31½in) l. 160cm (63in)
Sele 2, Switzerland

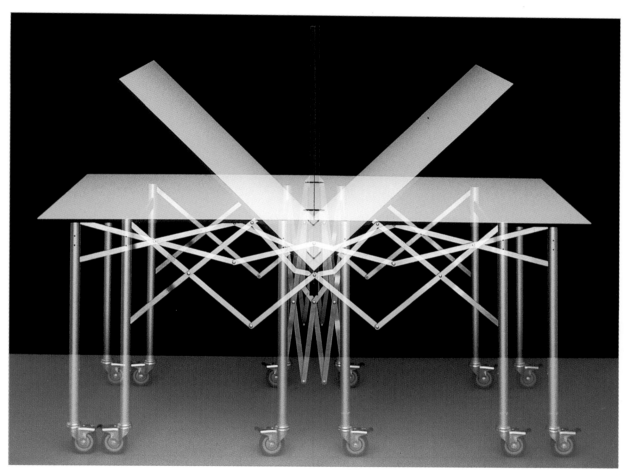

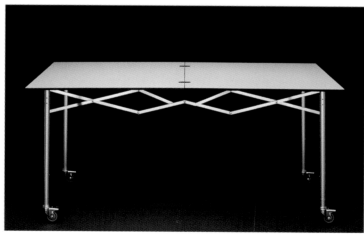

Pekka Tiovola
Electronically adjustable table, Promo
Laminates, veneer, steel
h. 70–120cm (27½–47¼in)
Martela OYJ, Finland

L. Carniatto
Workstation, Navigator
Aluminium, stratified laminate
h. 132cm (52in) l. 102cm (40⅛in) d. 44cm (17⅜in)
Bellato, Italy

Tim Power
Computer desk, Go-Car
Cut glass, metal
h. 81cm (31⅞in) w. 52cm (20½in) d. 72cm (28⅜in)
Zeritalia, Italy

Geoff Hollington
Home office desk, vuTable
Polyester, resin, steel
h. 130cm (51¹/₈in) w. 100cm (39³/₈in) d. 68cm (26³/₄in)
Herman Miller, UK (Sponsor)
Prototype

The vuTable has taken the idea of the home office to the nth degree. This is the interactive table for the family of the future where all communication (business, domestic and school work) is done through e-mail, voice and video accessed by speaking and moving hands. Even socializing can be achieved without leaving the chair. A built-in camera sends one's image to join friends and relatives at a virtual meeting table on the wide-angled screen.

Dante Donegani and Giovanni Lauda
Home office, Big Foot
Metal, wood
h. 72cm (28³⁄₈in) w. 75cm (29½in) l. 125–325cm (49¼–128in)
Radice SNC, Italy

Alberto Meda, Paolo Rizzatto
Office system, Partner Office
Beech, aluminium
h. 125cm (49in) w. 70cm (27½in) l. 100cm (39³⁄₈in)
Kartell SpA, Italy

Dante Donegani and Giovanni Lauda
Home office, Midi
Metal, wood
h. 80cm (31½in) w. 60cm (25⅜in) l. 135cm (53⅞in)
Radice SNC, Italy
Limited batch production

Dante Donegani and Giovanni Lauda
Home office, Compact
Metal, wood
h. 80cm (31½in) w. 60cm (25⅜in) l. 110cm (43⅜in)
Radice SNC, Italy
Limited batch production

Pepe Tanzi
Office system, Popoffice
Aluminium, laminates, resin
Album srl, Italy

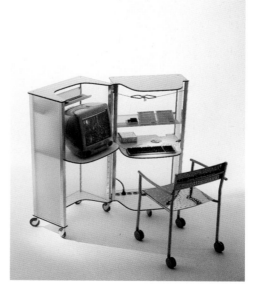

Teppo Asikainen
Wall Panel, Soundwave
Moulded polyester fibre
l. 60cm (23⅝in) h. 60cm (23⅝in)
Snowcrash, Sweden

Annette Egholm and Jacob Agger
Screen, Shade
Aluminium, woven textile
Large: h. 188cm (74in) w. 89cm (35in)
Small: h. 153cm (60¼in) w. 89cm (35in)
Bent Krogh A/S, Denmark

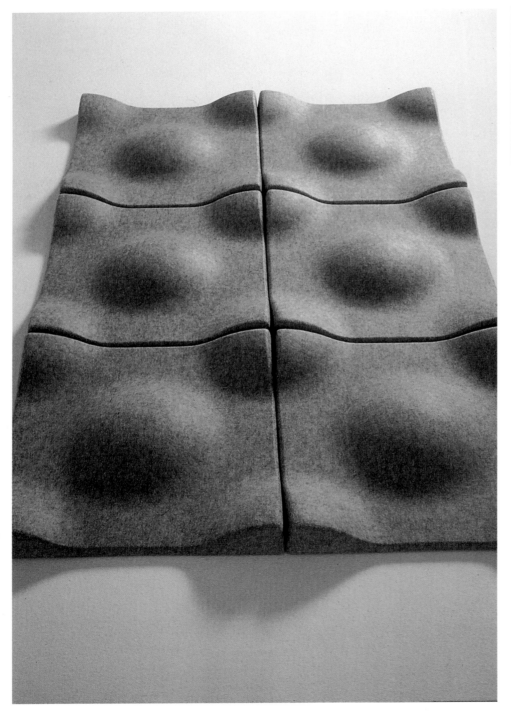

Snowcrash was founded in the late 1990s as an innovative group of young Finnish designers. It received international press coverage following its debut exhibition in Milan in 1997 which released into the market products which have become synonymous with the idea of exciting experimental design. Following this success, it was bought by a Swedish firm, Proventus Design, owners of Artek and the textile company Kinnasand. This year the group has embarked on a new stage of its development. Although recently Snowcrash has been less in evidence, the Proventus group has invested heavily in its development and it is now on course to expand upon its original vision.

Teppo Asikainen's 'Soundwave' wall panel, launched in Cologne as an early prototype and now in production, is made from moulded polyester fibre, a material used in the interior panels of aircraft, cars and trains. This felt-like polyester is composed from recycled plastic bottles and different fabrics can be added for varying finishes and colour combinations. Each hand-made tile is joined to the next with Velcro.

Antonio Citerio
Storage unit collection, Cross
Steel, oak, wenge, glass, aluminium, marble
Various sizes
B&B Italia, Italy

Hannes Wettstein
Shelves, Items
Wood, aluminium
h. 15/62/76cm (6/24/30in) l. 180/270cm (71/106in)
d. 64cm (25in)
Cassina SpA, Italy

Peter Maly
Furniture system, Duo-Medienbank
Maple, aluminium
Various sizes
Interlübke GmbH, Germany

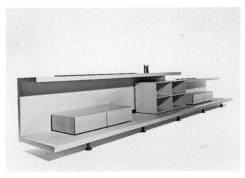

Lorenzo Damiani
Stereo and CD stand, Flex
Wood, steel
h. 100cm (39in) w. 150cm (59in) d. 50cm (19⁵⁄₈in)
Prototype

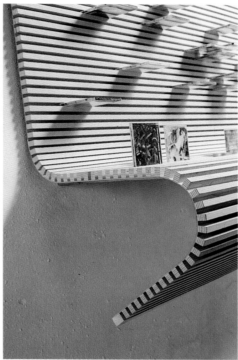

Gabriela Nahlikova, Leona Matejkova
Cabinet, Cube
PUR
h. 40cm (15¾in) w. 40cm (15¾in) l. 40cm (15¾in)
Prototype

Louise Campbell
Storage unit, Casual Cupboards
Veneer, ash, elastic, velcro
h. 180cm (70⅞in) w. 40cm (15¾in) d. 30cm (11⅞in)
Bahnsen Collection, Denmark

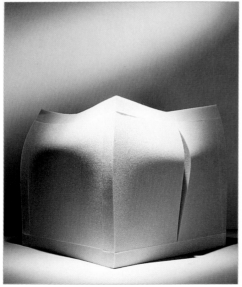

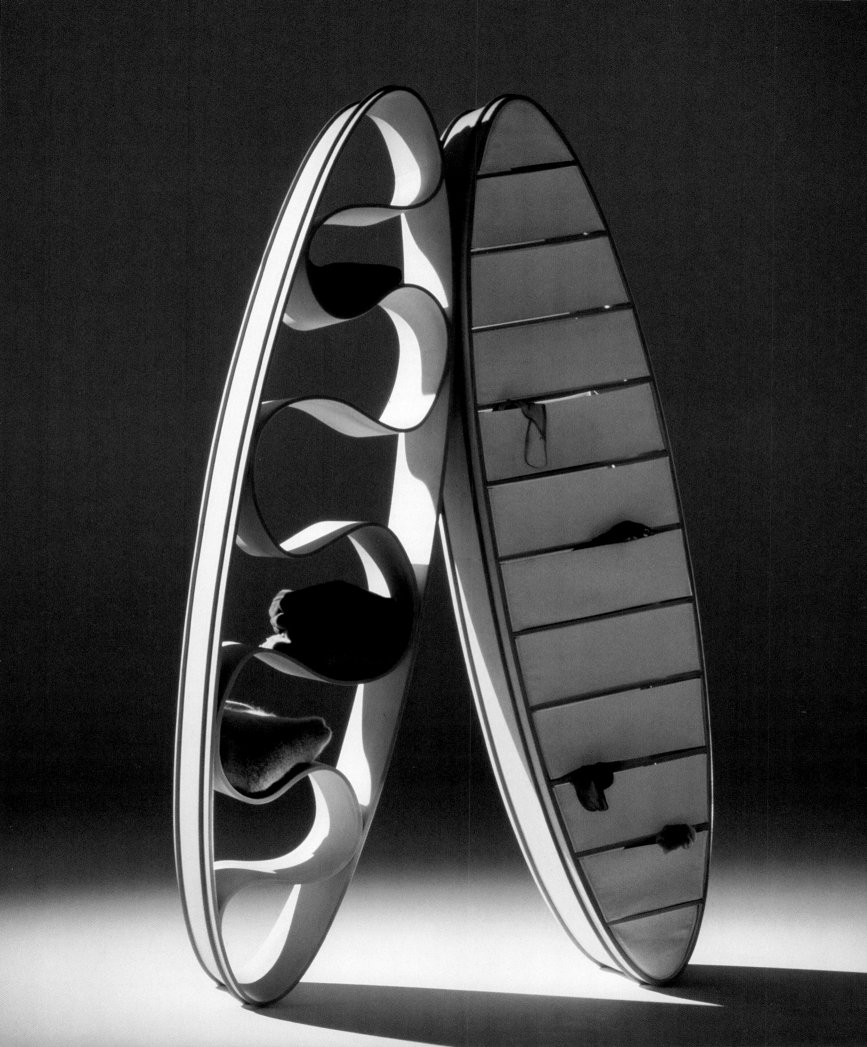

**Benny Mosimann
Storage system, Wogg 20
PET, veneered beech, aluminium
h. 96/184cm (37¾/72½in) w. 76cm (30in) d. 57cm (22½in)
Wogg AG, Switzerland**

Wogg 20 is a development of Benny Mosimann's range of sideboards and cupboards which share the same sliding, transparent fronts. '20' is a larger container with a more sophisticated and rounded skinned form coupled with a technologically innovative casing. A double PET construction which has a smooth exterior joined to a moulded inner base means that vertically the construction is extremely stiff but horizontally it is very flexible. The striped pattern, however, makes it appear rigid while its transparency maintains the delicacy of its appearance.

Hannes Rohringer
Card-Mobil
Wood, lacquer, white aluminium
h. 171cm (67³/₈in) w. 39cm (15³/₈in) l. 156cm (61³/₈in)
Streitner GmbH, Austria

Hannes Rohringer
Card Roadster range
Wood, lacquer, white aluminium
Streitner GmbH, Austria

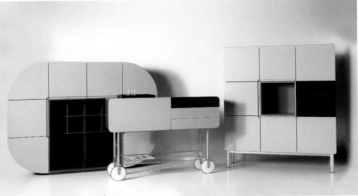

Stephan Titz
Korpus system, Homebase
Plum wood/beech
h. 130cm (51in) w. 130cm (51in) l. 130cm (51in)
Team 7, Austria
Limited batch production (Plum wood)
Mass production (Beech)

Stephan Titz's 'Homebase', Procter Rihl's 'Flow' and Jiri Pelcl's 'Pebble' and 'Erratic Block' (see p80) all challenge the two-dimensional appearance of the traditional bookcase which can seem formulaic. 'Homebase' is a series of square oblique cubes which can be twisted and teased as desired. Similarly, 'Flow' is completely flexible (the designers consider it to be about randomness, saying 'the user chooses the layout').

Fernando Rihl, Christopher Procter
Shelving, Flow
Birch plywood
h. 227.5cm (89½in) w. 227.5cm (89½in) d. 37.5cm (14¾in)
Spatial Interference, London
Limited batch production

Shigeru Uchida
Shelves, Kaja
Laminates
h. 350cm (138in) l. 210cm (82½in) d. 22cm (8½in)
Abet Laminati SpA, Italy

Ettore Sottsass
Furniture system, Kampa
Composite veneer, steel
h. 211cm (83in) w. 66cm (26in) l. 112cm (44⅛in)
Memphis srl, Italy

Jiri Pelcl
Seat-bookcase, Pebble
Wood, acrylic
h. 50cm (19⁵⁄₈in) l. 118cm (46½in) d. 48cm (19in)
Atelier Pelcl, Czech Republic

Jiri Pelcl
Bookcase, Erratic Block
Wood, acrylic
h. 130cm (51⅛in) w. 110cm (43⅜in) d. 50cm (19⅝in)
Atelier Pelcl, Czech Republic

Martine Bedin
Bookcase, Slate
Laminates
h. 180cm (71in) w. 50cm (19½in) l. 140cm (55in)
Abet Laminati SpA, Italy

Maarten Van Severen
Low Cupboard on Castors 94
Aluminium
h. 40cm (15⅞in) w. 240cm (94½in) d. 40cm (15⅞in)
Maarten Van Severen Meubelen, Belgium

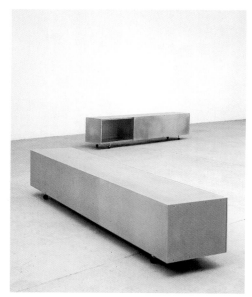

Alfredo Häberli, Christophe Marchand
Modular structure, SEC
Aluminium, steel, glass, methacrylate
Various sizes
Alias srl, Italy

Dakota Jackson
Bookcase, Mainframe
Lacquer, anigre, glass, acrylic
h. 219.7cm (86½in) w. 115cm (45¼in) d. 38cm (15in)
Dakota Jackson, USA

Carlo Cumini
Storage system, Cut-al
Aluminium
h. 34–194cm (13³/₈–76³/₈in) l. 64–328cm (25¹/₈–129¹/₈in)
d. 49cm (19¼in)
Horm srl, Italy

Studio Technico Horm
Cabinet, Expò
Aluminium, glass
h. 194cm (76³/₈in) l. 32–64cm (12⁵/₈–25¹/₈in) d. 34cm (13³/₈in)
Horm srl, Italy

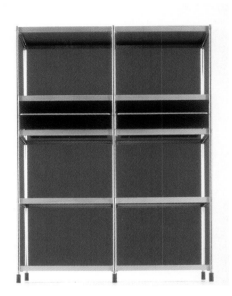

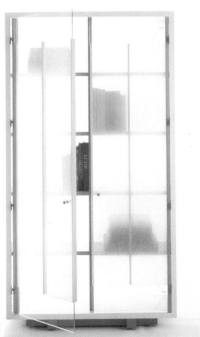

Piero Lissoni
Cabinet, One
Metal, lacquer, macroter
h. 64cm (25⅞in) l. 270cm (106¼in) w. 61.2cm (24⅞in)
Cappellini SpA, Italy

Carlo Tamborini
Furniture system
Stainless steel, Lacquer/oak
h. 115cm (45¼in) w. 254cm (100in) d. 66cm (26in)
Pallucco Italia, Italy

Petra Runge
Mirror, Book
Anodized aluminium
h. 127cm (50in) l. 47/90cm (18½/35½in)
De Padova srl, Italy

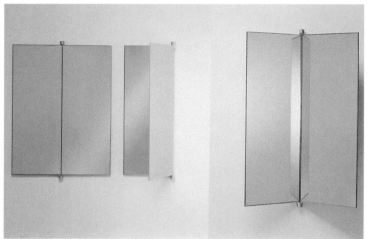

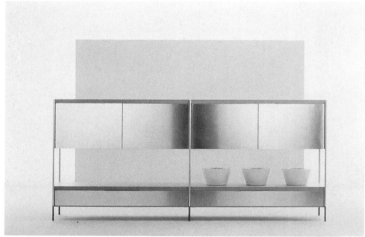

Marco Giunta
Office Furniture, L4RU
Cardboard
h. 188cm (74in) w. 36cm (14⅛in) l. 30cm (11¾in)
Disegni, Italy

Martin Szekely
Cupboard, Armoire
Aluminium, plastic
h. 107cm (42⅛in) w. 64cm (25⅛in) d. 41cm (16⅛in)
Galerie Kreo, France
Limited batch production

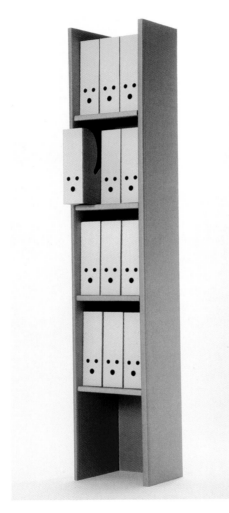

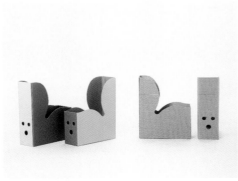
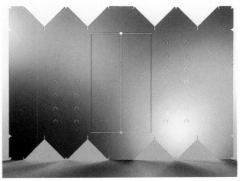

Jurgen Bey
Chair, do add #1
Laminate, chromium plated steel
h. 83cm (32⅝in) w. 40cm (15¾in) d. 45cm (17¾in)
do + Droog design

Marti Guixé
do reincarnate
Nylon thread, fitting
do + Droog Design

Marti Guixé
Lamp, do scratch
Light box, black coating
h. 9cm (3½in) w. 65cm (25¼in) d. 9cm (3½in)
do + Droog Design

Droog Design was founded in 1993 and is based in Amsterdam. It initiates and develops experimental projects from a group of young designers, each collection reflecting an ideology. In 1999 its 'Couleur Locale' for Oranienbaum provided a coherent proposition for the regional identity of a currently impoverished area in former East Germany.

In 2000, 'do Create' has brought together designs from ten individuals or studios – which include Radi Design (France), Marti Guixé (Spain), Thomas Bernstrand (Sweden) and Jurgen Bey (The Netherlands). They have invented a series of designs for the 'do' experimental brand which was set up by the Dutch publicity firm Kesselskramer. By asking for products to fit a brand name rather than creating a brand from existing items, Kesselskramer offered a clear canvas for invention and improvization. '"do", as the name suggests, is an ever-changing brand that depends on what you do. A brand that is open to ideas from anyone and anywhere.' Droog has created objects with which users interact emotionally and physically. This concept is not new to Droog. Gijs Bakker's 'Peep Wallpaper' works only when combined with the old paper or posters it covers, and Dyoke de Jong's curtains, complete with jacket pattern, can be turned into an item of clothing when no longer needed. However, the 'do' collection gives a new slant to the idea of personal production. Users influence the design and the design itself becomes an indication of their character.

do add 1
In order to work, one has to add to the chair. What appears to be a broken chair turns into something brand new, simply by thinking and doing, says the catalogue.

do reincarnate
Any tired or over-familiar possession can be reanimated. One can slip the invisible thread around any old lamp or painting, attach via a light cable and let it hang from the ceiling as an exciting new luminaire.

do scratch
No light shines from this lamp unless one does something about it. By scratching one's own message or graffiti a personalized light source emerges.

Dinie Besems, Thomas Widdershoven
String of chains, do connect
Silver, diverse metal, plastic
l. 13cm (5⅛in) per piece
do + Droog Design

Frank Tjepkema, Peter van der Jagt
Vase, do break
Porcelain, rubber, silicone
h. 34cm (13³⁄₈in) di. 15cm (6in)
do + Droog Design

Jurgen Bey
Bench, do add #2
Laminate, chromium plated steel
h. 83cm (32⅝in) w. 110cm (43½in) d. 45cm (17¾in)
do + Droog Design

Marijn van der Poll
Armchair, do hit
1.25mm steel
h. 75cm (29½in) w. 100cm (39³⁄₈in) d. 70cm (27½in)
do + Droog Design

Thomas Bernstrand
Light, do swing
Stainless steel, lampshades
do + Droog Design

Marti Guixé
Frame, do frame
Self adhesive tape
w. 5cm (2in) l. 100m (327 ft)
do + Droog Design

do connect
The potential of this string of 13 chains is endless – why not design jewellery, sink or dog lead?

do break
Not even the worst argument can spoil the beauty of this vase. No matter how hard one throws it, although the exterior will show the vestiges of aggression, the vase will remain intact thanks to the sticky rubber interior.

do add 2
Play with this bench to discover its absurd functionality. Only by adding to the opposite end can one make the design balance. Anything can be used, from a trusting friend, Alsatian dog, or, more daringly, something hot or breakable.

do hit
Be the co-designer of an armchair. Take the hammer and the steel cube and bash it into any shape you wish.

do swing
The living room can become playroom or gym. By suspending the lamp, which is fitted with two light bulbs and shades, from the ceiling one can swing away all day.

do frame
A simple roll of ornate printed tape can turn any picture, no matter how banal, into a spectacular work of art.

Lighting

This year has seen the breaking of the speed of light. Dr Lijun Wang of the NEC Research Institute at Princeton University has made a pulse of light exist in two places at once by making it travel 300 times faster than the speed of light. Part of Einstein's theory of relativity has been challenged and an opportunity has opened up to better understand the nature of light and how it behaves. Although not quite as earth-shatteringly important, a revolution is also happening in the world of lighting design.

The use of Light Emitting Diodes (LEDs) in ambient lighting is starting to hit the marketplace. The commercial use of LEDs in the lighting industry is still in its infancy. Previously used in the rear lights of cars or as control lamps in electronic items, their development for use in luminaires is already changing the way we conceive light tools. At the moment, they are very expensive, as many LEDs are needed to illuminate a space, but their low temperature, focused light, low maintenance and long life (15 years) guarantee their development. LEDs are the size of pin heads and are based on semiconductor compounds which convert electricity directly into light.

Siemens has succeeded in lighting a room with LEDs. Although red, green and orange LEDs have been available for some time, the blue version has recently been developed. Siemens has mixed this 'new' blue light with the existing colours to obtain an effect comparable to the light given off by a traditional bulb; it has then integrated over 14,000 of these LEDs into the ceiling of an installation created for the Tirol Architectural Forum in Innsbruck. In addition, Siemens has developed O.L.E.D. (Organic Light Emitting Diode) which is a high-tech foil, only 1mm thick. A 16-square-centimetre panel has been developed which can not only produce extremely bright light but can also handle video signals making the idea of a wallpaper that suddenly changes colour and mutates into a television screen not just a science fiction possibility but a reality.

Michele De Lucchi has selected four examples of LED technology: Snowcrash's development of the Globlow lamp, Zumtobel's 'Phaos' and 'Ledos' and Ingo Maurer's prototype 'Bellissima Bruta'. There were few examples to choose from but we will be expecting to see more next year.

Another important development is in office lighting where a new standard dictates that luminance has now been restricted to 200 candles per square metre, shining at an angle of 60 degrees from the vertical to secure optimum visual comfort for workstations. Among examples selected are Gecchelin's Light Air System, Hosoe's Onda, Arca and Vola lights and Gismondi's Megan System. De Lucchi also recognized the number of lights which have both direct and indirect light sources, making them suitable for either the office or home. Notable among these were examples by Tobias Grau, King and Miranda and Claudio Bellini.

On a more relaxed note, we have also featured lights made out of plastic bottles ('Alpha' by Sigi Bussinger and Iwan Seiko); plastic cups (Mo-billy's 'Cuplight'); light bulbs ('Follow Your Bliss' by Ralph Ball); a plastic hat (Paolo Ulian's 'Palombella') and have paid tribute to the poetry of Out Design's 'Flow'. Witty puns on form and function can be seen in the light which stands on its own cord ('Flapflap' by Büro für form), the Wallpaper Lamp by Droog's Jaap van Aarkel and Dumoffice's 'Whoosh'.

Light affects our emotions and lighting design is constantly evolving as new technologies appear. Each year, more designers not normally associated with lighting seem to experiment with the medium and the selection represents the wealth of designs on offer.

Claudio Bellini
Suspension lamp, ITI
Glass, crystal
100w frosted halogen bulb
h. 100–200cm (39³/₈–78⁷/₈in) di. 40cm (15³/₄in)
Artemide, Italy

Architettura Laboratorio
Suspension lamp, Saturno
Aluminium
55w fluorescent bulb
h. 100–160cm (39³/₈in)
di. 58cm (22⁷/₈in)
Artemide, Italy

Artemide has produced a collection of luminaires suited equally to the home or office environment. Claudio Bellini's glass suspension lamp, ITI, comprises two diffusers: the outer in satinized crystal and the inner in painted glass. A lever enables the lamp to be moved inside the diffuser as required for different lighting effects. Similarly, Architettura Laboratorio's suspension lamp, 'Saturno' in painted aluminium has a double soft light emission which makes it particularly suitable to domestic environments as well as for VDU terminals.

Dumoffice
Wall/ceiling suspended lamp, Surve
Artglass
Fluorescent bulb
l. 75cm (29½in) w. 22cm (8½in)
Belux AG, Switzerland

Dumoffice
Hanging light, Whoosh
Handblown frosted glass
2 x 40w bulb
w. 38cm (15in) h. 20cm (7¾in) d. 7.5cm (3in)
Dumoffice, The Netherlands
Prototype

Dumoffice was founded in 1997, three years after its creators, Wiebe Boonstra, Martijn Hoogendijk and Marc van Nederpelt graduated from The Design Academy of Eindhoven, The Netherlands. Their work is characterized by a fresh and witty approach to the everyday object which has crossed the boundaries from the limited batch production of their earlier pieces to the commercial success of work now being mass produced by Belux and Hidden of sdb Industries. The 'Whoosh' swinging light and 'Surve' armature for two strip lights, illustrate this duality. The former is an act of creation, designed as a frozen moment of movement. It refers to the photos of Pablo Picasso making virtual drawings with a torch. 'Surve', on the other hand, while retaining 'the spark of a luminous idea' (Robert Thiemann, *Frame Magazine*), is a highly-engineered and technically-conceived item of industrial design.

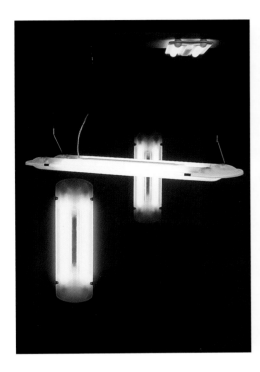

Mauro Marzollo
Lighting fixture/hanging lamp, Volo
Glass, metal
3 x 60w bulb
h. 47cm (18½in) di. 38cm (15in)
Murano Due, Italy

Tobias Grau
Suspension lamp, Oh China
Bone china
60w halogen bulb
Adjustable height, di. 9cm (3½in)
Tobias Grau, Germany

Tobias Grau
Suspension lamp, Project X
Aluminium, PC glass
150w halogen bulb
Adjustable height, di. 40cm (15¾in)
Tobias Grau, Germany

Tobias Grau's 'Project X' is a height-adjustable hanging lamp with a dual light effect packed into a simple shape. Intense non-dazzling light is released downwards through opal glass while at the same time a soft light shines upwards through a coloured glass diffuser. Once again, a major designer/manufacturer is addressing the need for dual purpose luminaires, suitable both for work and relaxation.

Marco Carenini
Ceiling lamp, Lubia (3 different lampshades)
Polypropylene
h. 23cm (9in) di. 6cm (2in)
Globe Energy Saver 21w
Prototype

Isao Hosoe
Lamp, Vola
Acidized blown glass,
chromed metal,
polycarbonate
75w halogen bulb
h. 7–10cm (2³⁄₄–3⁷⁄₈in)
l. 24cm (9¹⁄₂in)
Luxo Italiana, Italy

Isao Hosoe, Peter Solomon
Lamp, Onda
Aluminium, microperforated sheet metal, polycarbonate
2 x 54w HFG5
h. 1.6cm (⁵⁄₈in) w. 50cm (19⁵⁄₈in) l. 170cm (67in)
Luxo Italiana, Italy

Makoto Kawamoto
Lamp, Frozen
Polycarbonate, zinc plated aluminium, PVC
40w e-14 bulb
h. 22.5cm (9in) w. 21cm (8¹⁄₄in) d. 21cm (8¹⁄₄in)
Aliantedzione, Italy
Limited batch production

Isao Hosoe, Peter Solomon
Office Lamp, Arca Dome Suspension
Aluminium, ABS plastic, polypropylene, steel
2 x 55w fluorescent bulbs
h. 10–60cm (4–23⁵⁄₈in) l. 120cm (47¹⁄₄in)
Luxo Italiana, Italy

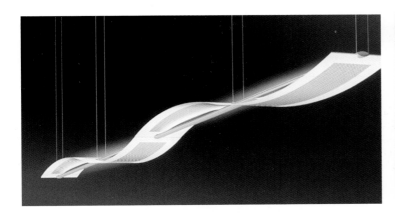

Sigi Bussinger and Iwan Seiko
Light installation, Alpha
1,000 plastic bottles
150w powerstar
di. 180cm (70⁷⁄₈in)
One-off

Henrik Kjellberg and Mattias Lindqvist
Pendant Lamp, IKEA PS
Chrome plated steel, polycarbonate, polypropylene
h. 48cm (19in) di. 35cm (14in)
IKEA, Sweden

Ralph Ball
Lamp, Very Light Box
Plate steel, assorted lamps
4 x 60w bulb
h. 25cm (9⁷⁄₈in) w. 25cm (9⁷⁄₈in) l. 32cm (12¹⁄₂in)
Ligne Roset, France

Ralph Ball
Lamp, Golden Delicious
Metacrylate moulded bowl and light bulbs
60w bulb
h. 20cm (7⁷⁄₈in) di. 30cm (11⁷⁄₈in)
Ligne Roset, France

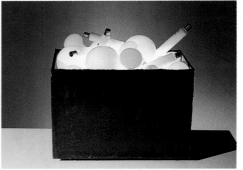

Fabrice Berreux
Lamp, Watt Colonne
Lacquered metal base, light bulbs
9 x 25w bulb
h. 190cm (74³/₄in) di. 28cm (11in) base 30 x 30cm (11⁷/₈ x 11⁷/₈in)
dix heures dix, France

Neil Austin
Light, Cuplight
Vending cups, copper wire frame
di. 60cm (23⁵/₈in)
Mo-billy, UK
Limited batch production

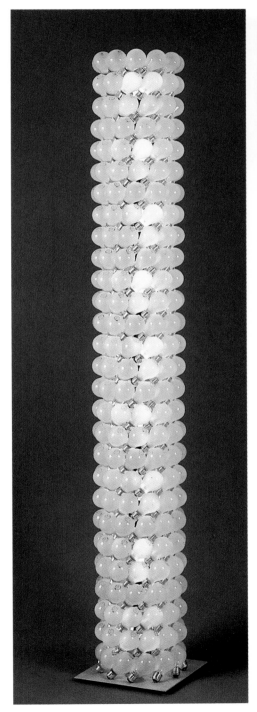

Ingo Maurer
Light, Bellissima Bruta
Printed circuit board, LED, steel, 40w bulb
h. 50cm (18⁷⁄₈in)
Ingo Maurer GmbH, Germany

Knuth Eckhard, Ingo Maurer
Hologram light, Holonzki
Glass, metal, stainless steel, 35w bulb
h. 18cm (7in) w. 13cm (5in)
Ingo Maurer GmbH, Germany

Ingo Maurer
Lighting fixture, Pierre ou Paul
Aluminium, steel, gold/platinum, 300w bulb
di. 100–120cm (39³⁄₈in–47in)
Ingo Maurer GmbH, Germany

Ingo Maurer
Light, Red Ribbon
Aluminium
Halogen bulb
l. 400cm (157½in)
Ingo Maurer GmbH, Germany
One-off

Ingo Maurer
Light, Yaki Mei
Paper, steel, silver, ventilators
l. 700cm (275½in) di. 80cm (31½in)
Ingo Maurer GmbH, Germany
One-off

Ingo Maurer's hauntingly beautiful 'vase of flowers', 'Bellissima Bruta', creates poetry from cutting-edge technology. Red and green LEDs have been combined with blue to produce a soft white light.

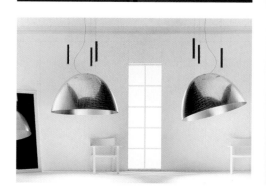
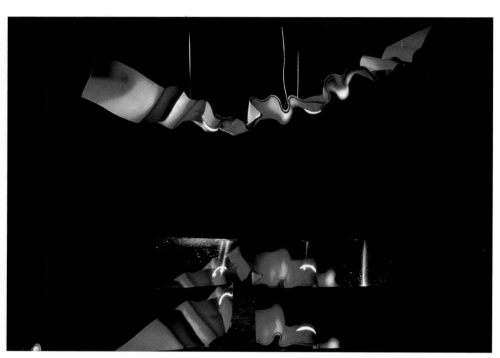
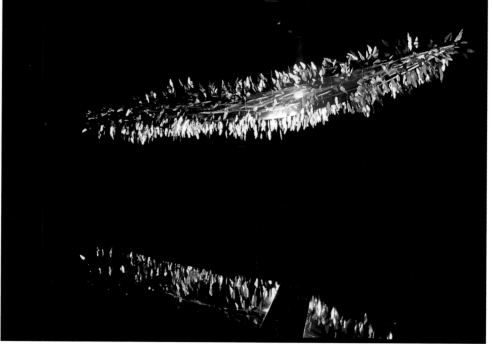

Ingo Maurer
Light, Max Mover
Aluminium, fibreglass
Max 100w bulb
Radius 150cm (56⁵⁄₈in)
Ingo Maurer GmbH, Germany

Matteo Thun
Lamp, Sphera T20
Glass, wood, chrome
60w bulb
Large: h. 61cm (24in)
Small: h. 42cm (16½in)
Leucos, Italy

Karim Rashid
Ceiling lamp, Kovac Lamp
Hand-blown Murano glass, chrome plated steel
75w bulb
h. various w. 40.5cm (16in)
Karim Rashid Industrial Design, USA

Karim Rashid
Table lamp, Kovac Lamp
Hand-blown Murano glass, chrome plated steel
75w bulb
h. 45.7cm (18in) w. 40.5cm (16in)
Karim Rashid Industrial Design, USA

Arik Levy
Light, Rewindable light
Snowcrash, Sweden

Vesa Hinkola, Markus Nevalainen, Rane Vaskivuori
Lamp, Globlow LED
Steel, rip stop nylon
Snowcrash, Sweden

Globlow LED is a development of the original Globlow lamps produced for Snowcrash. Its designers, Vesa Hinkola, Markus Nevalainen and Rane Vaskivuori, wanted to see if the concept of living light could progress further. The original lamp inflated when it was switched on, its sequel uses LED and microprocessing technology to invent an interactive light which can be programmed by computer to play different light sequences. It can be activated by remote control or by mobile phone and eventually one will be able to download different programmes from the Snowcrash website.

Marc Krusin
Light, Bulbed Wire
Polycarbonate, aluminium, electrical cable
8 x 15w Osram Nitra bulbs
h. 120–180cm (47–71in) di. 70cm (27½in)
Limited batch production

Philippe Starck
Chandeliers, Cicatrices des Lux 3, 5 and 8
Hand ground crystal, polished crystal sheets, conductive varnish
Flos, Italy

Cicatrices des Lux 3
3 x max 35w GY6.35 HS
h. 80cm (31½in), slab 32 x 32cm (12⅝in x 12⅝in)

Cicatrices des Lux 5
5 x max 35w GY6.35 HS
h. 80cm (31½in), slab 87 x 19cm (34¼in x 7½in)

Cicatrices des Lux 8
8 x max 35w GY6.35 HS
h. 120cm (47¼in) slabs 43.5 x 43.5cm (17⅛in x 17⅛in) / 32 x 32cm (12⅝in x 12⅝in) /
17 x 17cm (6⅞in x 6⅞in)

The Borderlight show in Milan brought together a group of young designers, some of them first time exhibitors, to create a collection of lamps which were characterized by their simple lines and lack of ostentatious decoration. The objective was to produce designs which relied on their structure and the material used to give birth to new and surprising light fittings and effects. Marc Krusin's 'Bulbed Wire' is a variation on the domestic chandelier, however, the fact that it is fabricated from cable sheathing and light bulbs held by a circle of glass transforms a design staple into an entirely new concept.

'Life is all about a continuum of passions and scars. The Cicatrices des Lux collection sets out to capture its surrealistic, crystalline and luminous nature' – so goes Starck's explanation of his new design for Flos.

Taco Langius
Lamp, Yuck 2
Gel, PVC
Flexneon
h. 4cm (1½in) w. 30cm (11¾in) l. 60cm (23⅝in)
Codice 31, Italy
Limited batch production

Andrea Branzi
Lamp, Cactus
Bronze, stainless steel, blown glass
60w bulb
h. 50cm (19⅝in) d. 32cm (12⅝in)
Design Gallery Milano, Italy

Marco Carenini
Lamp, Jim
Acrylic
h. 31cm (12in) di. 21cm (8in)
Globe Energy Saver 21w
Prototype

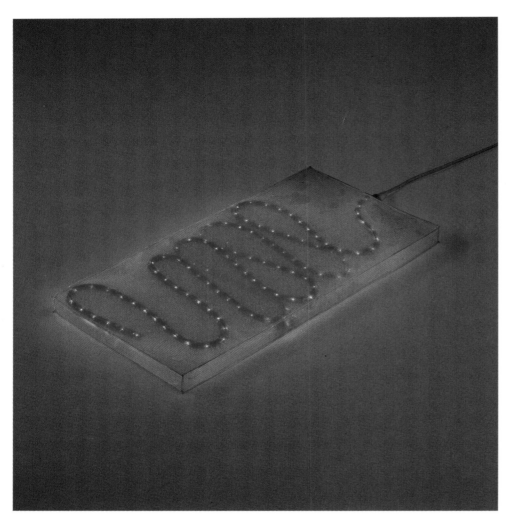

Valter Bahcivanji
Lamp, T Kiluz
Steel, polypropylene
60w bulb
h. 25cm (10in) di. 13cm (5in)
Agora e Moda Ind. Com. Ltda, Brazil

Valter Bahcivanji
Floor lamp, Two Points
Polypropylene, stainless steel
40w bulb
h. 180cm (70⁷/₈in) d. 28cm (11¹/₈in)
Agora e Moda Ind. Com. Ltda, Brazil

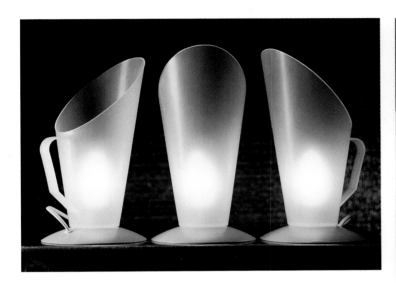

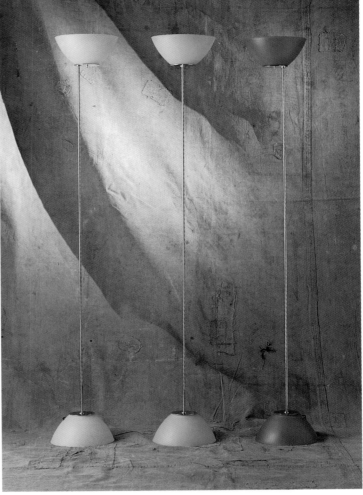

Sergio Brioschi
Lamp, Nomade
Polycarbonate, ABS
50w/60w fluorescent cold cathode
di. 9/10cm (3½/3⅞in), varying lengths
Antonangeli Illuminazione, Italy

Jaap van Aarkel
Lamp, Wallpaper Lamp
Wallpaper, resin
h. 42cm (16½in) w. 13cm (5⅛in) d. 11cm (4⅜in)
Droog Design, The Netherlands
Prototype

Büro für form
Lamp, Flapflap
Iron, plastic
25–40w bulb
h. 50/80cm (19⅝/31⅞in)
Next Design GmbH, Germany

Jaap van Aarkel's 'Wallpaper Lamp' for Droog Design plays on the idea of subtraction and addition. The part of the wallpaper used for the lampshade creates a void on the wall behind.

Bernard Brousse
Lamp, Ellipsis
Polyurethane, technopolymer
Max 20WE 27 Fluo E1
h. 29.5cm (11⁵/₈in) w. 48cm (18³/₄in)
Baleri Italia SpA, Italy

Toni Cordero
Floor lamp, Nuvola
Micro-holed, chromium-plated standard,
black Marquinia marble base
Max 300w + max 54w bulb
h. 220cm (86⅝in) l. 35cm (13¾in) d. 30cm (11¾in)
oluce, Italy

Toni Cordero
Wall light, Nuvola
Anodized metal
Max 3 x 100w bulbs
l. 80cm (31½in) w. 50cm (19⅝in) d. 18cm (7in)
oluce, Italy

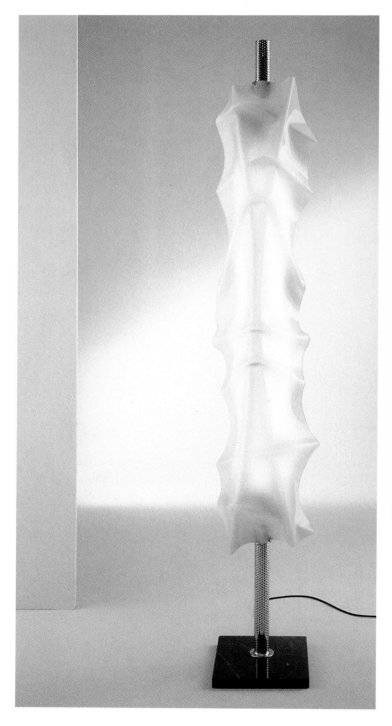
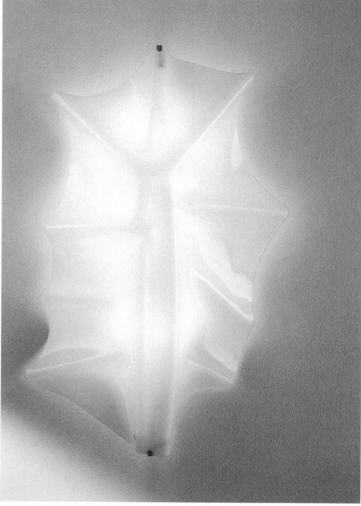

Tsutomu Kurokawa
Lamps, Flow series
Polycarbonate
Table lamp: h. 43cm (17in) w. 22.5cm (8⁷/₈in)
Suspension lamp: h. 21cm (8¹/₄in) w. 22.5cm (8⁷/₈in)
Bracket light: h. 32.5cm (12³/₄in) w. 16.5cm (6¹/₂in) d. 21cm (8¹/₄in)
Daiko Electric Co. Ltd, Japan

Out DeSign has been formed by Tsutomu Kurokawa following his split from Masamichi Katayama and the disbanding of H-Design. His 'Flow' series of lights were of particular interest to Michele De Lucchi who admired their poetic qualities. The design was influenced by the beauty Kurokawa discovered in a view of a lamp reflected continuously between a series of mirrors. This led him to research the idea of reflecting lights within a single lighting tool. Double layers of shades made from half mirrors shine refracted light on the wall and ceiling. The result is a transparent space, a light experience rather than a simple lamp.

Paolo Ulian
Table lamp, Palombella
Steel, silicon rubber
Energy saving bulb
h. 50cm (19⁵/₈in) w. 35cm (13³/₄in) d. 8cm (3¹/₈in)
Prototype

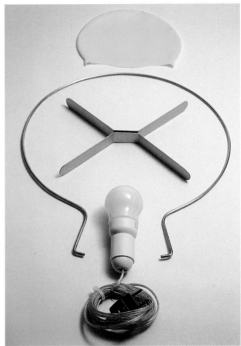

Arik Levy
Light, Alchemy
Glass, metal
20w halogen bi-pin
h. 18cm (7in) di. 13cm (5⅛in)
Tronconi, Italy

Fabrice Berreux
Lamp, Pise
Four lampshades, lacquered metal base
4 x 60w bulbs
h. 200cm (78⁷⁄₈in) base: 48 x 32cm (18⁷⁄₈ x 12⁵⁄₈in)
dix heures dix, France

Steven Holl
Wall lamp, Kiasma
Aluminium
150w R7s halogen bulb
w. 134cm (52³⁄₄in) h. 12cm (4³⁄₄in) d. 12cm (4³⁄₄in)
FontanaArte, Italy

Steven Holl
Wall lamp, Triple
Aluminium, glass
3 x 60w E14 bulbs
h. 55cm (21³⁄₈in) w. 43cm (17in) d. 6cm (2³⁄₈in)
FontanaArte, Italy

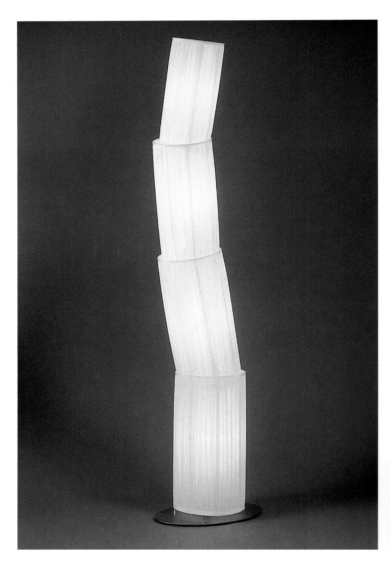

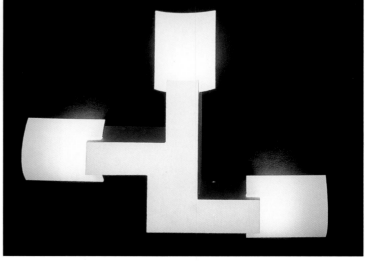

Maurizio Peregalli
Floor lamp, Star Cube Terra
Steel, polycarbonate
250w bulb
h. 176cm (69¼in) w. 22cm (8⅝in) l. 22cm (8⅝in)
Zeus, Italy

Jorge Garcia Garay
Light, Isis
Acrylic
60w bulb
h. 16cm (6¼in) w. 8cm (3⅛in) d. 14cm (5½in)
Garcia Garay SL, Argentina

Jorge Garcia Garay
Library light, Libra
Acrylic
60w bulb
h. 16cm (6¼in) w. 8cm (3⅛in) d. 14cm (5½in)
Garcia Garay SL, Argentina

Kazuhiro Yamanaka
Wall light, What a little Moonlight can do
Alcobond
12v halogen light
h. 120cm (47in) w. 158cm (62in) d. 4mm (⅛in)
Limited batch production

Jorge Garcia Garay
Wall light, Embassy
Iron, metacrylic
2 x 36w fluorescent bulb
h. 13cm (5⅛in) l. 132cm (52in) d.11cm (4⅜in)
Garcia Garay SL, Argentina

Johanna Grawunder
Light, F6 (Fractals Collection)
Aluminium
30w fluorescent bulb
h. 18cm (7in) w. 18cm (7in) l. 120cm (47¼in)
Memphis, Italy

Jan van Lierde
Architectural lighting, Secret
Steel, die cast aluminium
Dulux L 1 x 18w 1 x 55w
w. 30cm (11¾in) l. 15cm (6in)
Kreon NV, Belgium

Luc Vincent
Lighting fixture, Square Moon
Aluminium, polycarbonate
4 x 18w TC-L Lamps
h. 5.5cm (2⅛in) w. 24.2cm (9½in) l. 68.1cm (26¾in)
Modular Lighting Instruments, Belgium

Ernesto Gismondi
Table lamp, E-light
Polycarbonate, nylon, brass, aluminium
Microlight, 3w lamp
h. 32.5cm (12⅞in) l. 57.5cm (22⅝in)
Artemide SpA, Italy

Ernesto Gismondi
Wall light, Megan Wall
Steel
2 x 14–35w fluorescent bulb
l. 58.3–148.3cm (23–58⅜in)
Artemide SpA, Italy

Ernesto Gismondi
Ceiling light, Megan Suspension
Steel
2 x 28–54w fluorescent bulb
l. 59cm (23¼in)
Artemide SpA, Italy

Ernesto Gismondi
Floor lamp, Megan Floor
Aluminium, steel, polycarbonate
4 x 55w fluorescent bulb
h. 185cm (72⅞in) w. 59cm (23¼in) d. 39cm (15⅜in)
base: 48cm (18⅞in)
Artemide SpA, Italy

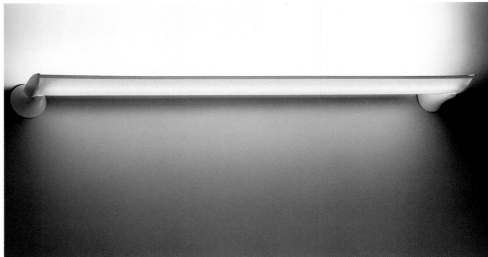

Ernesto Gismondi's (founder and Managing Director of Artemide) 'e-light' desk lamp belongs to a category of low energy consumption systems. Its microlight technology produces a lamp which uses very little electricity. At a time when we are once again being warned of the dangers of abusing our consumption of power, it is reassuring that such concerns form part of the design philosophy of a leading Italian light manufacturer. As Michael Horsham, General Editor of the *International Design Yearbook 1999*, wrote, 'Domestic lighting which does not draw upon the national grid – in whichever nation, is a project which designers, manufacturers and engineers should look at with the same application as they have the energy-saving bulb'. Michele De Lucchi confesses to 'a few big clouds on the horizon'. His concerns for the future are bound to the issues of pollution and to the sustainability of development on the planet. 'I think we must make an effort to be optimistic, with a deep awareness, however, that certain things need to be addressed without further delay' (*Domus*, October 1999).

Bruno Gecchelin
Lighting System, Light Air
Sheet steel
28w/54w fluorescent bulb
h. 4cm (1½in) w. 24cm (9½in) l. 170cm (68in)
Iguzzini Illuminazione, Italy

Felice Dittli
Ceiling Light, Box
Aluminium, glass
2 x 14w fluorescent
h. 20–50cm (7⅞–19⅝in) w. 47–120cm (18½–47¼in)
l. 96.9cm (38⅛in)
Regent, Switzerland

Felice Dittli
Ceiling Light, Tool
Aluminium
36/58w bulb
h. 3–7.6cm (1⅛–3in) w. 1.5–3.8cm (½–1½in)
l. 50.3–127.8cm (19⅞–50¼in)
Regent, Switzerland

Significant at Euroluce in 2000 were the number of VDU-friendly lighting systems. Bruno Gecchelin's 'Light Air', the 'Megan' collection by Gismondi and Isao Hosoe's 'Arca' (see p93) all use 'darklight'. Glare has been eliminated from office lighting to secure optimum visual comfort for workstations. Of the above systems 'Arca' offers the domestic environment these technological advantages by combining indirect and controlled direct light in one source.

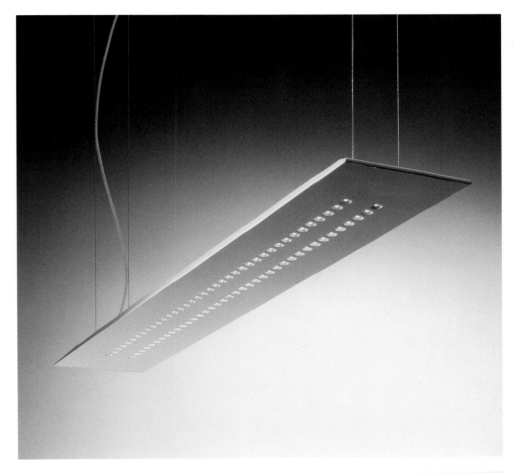

James Irvine
Suspension lamp, Float Rectangular
Aluminium, polycarbonate
39w fluorescent bulb
h. 100–160cm (39³⁄₈–63in) w. 103cm (40½in)
Artemide, Italy

James Irvine
Suspension lamp, Float Circular
Aluminium, polycarbonate
39w fluorescent bulb
h. 100–160cm (39³⁄₈–63in) w. 103cm (40½in)
Artemide, Italy

Lievore, Altherr and Molina
Ceiling light, Hopper 60–80
Aluminium, glass
2 x 150/250w halogen bulb
h. 210cm (82⅝in) w. 60–80cm (23⅝–31½in)
Metalarte, Spain

Christophe Pillet
Ceiling light, Easy Mechanics Sky
Painted metal, chromed metal
150w incandescent, 20w fluorescent, 250w halogen
h. 27.5cm (10³/₄in) di. 55cm (21⁵/₈in)
Tronconi, Italy

Tronconi was established in 1956 as a traditional lighting manufacturer. It wasn't until the 1970s that it started its collaboration with invited designers. Today, Fabrizio Tronconi (b. 1964) heads the company. He has started working with a number of young designers to produce a collection of simple and practical luminaires fully informed with new requirements regarding form and technology. 'Easy Mechanics Sky' by Christophe Pillet is a hanging lamp giving off diffused light. It is height adjustable through the counterweight which pierces the diffuser.

Antonio Citterio and Oliver Low
Light, Plaza
Die cast aluminium, prismatic glass
Max 300w R75 HDG incandescent bulb
di. 56cm (22in) d. 11.5cm (4½in)
Flos, Italy

Konstantin Grcic
Ceiling light, Hertz
Aluminium, techno-polymer, polycarbonate
Halogen bulb
h. 30cm (11⅞in) w. 35.5cm (14in) l. 35.5cm (14in)
Flos, Italy

Jasper Morrison
Light '01' and '02'
Iron, aluminium, polycarbonate
Flos, Italy

'01'
Max 100w E27 1AA incandescent bulb,
2 x 18w 2911 FSD fluorescent bulb
di. 38cm (15in) d. 10cm (4in)

'02'
Max 60w E14 IBP incandescent bulb,
13w G249 1FSQ fluorescent bulb
di. 30cm (11⅞in) d. 7.3cm (2⅞in)

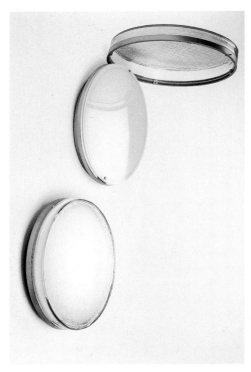

Asahara Sigeaki
Lighting, Karma IM
Die cast aluminium
70w/150w metal halide bulb
h. 32cm (12⅝in) w. 33cm (13in)
Lucitalia, Italy

Asahara Sigeaki
Lamp, Lxul
Die cast aluminium, blown glass
Max 150w incandescent/23w fluorescent
h. 25cm (9⅞in) w. 25cm (9⅞in) di. 17cm (6⅝in)
Lucitalia, Italy

Ross Lovegrove
Lamp, Agaricon
Polycarbonate with silk opaque treatment
150w bulb
h. 28cm (11in) di. 40cm (15¾in) base 8.5cm (3⅜in)
LucePlan, Italy

Dominique Perrault and Gaëlle Lauriot-Prevost
Floor lamp, M.A.
Steel, stainless net
300w R7s halogen bulb
h. 240cm (94½in)
FontanaArte, Italy

Andrea Branzi
Floor lamp, Anfora
Bronze, marble, blown glass
100w halogen bulb
h. 250cm (98⅜in) d. 42cm (16½in)
Design Gallery Milano, Italy

Andrea Branzi
Floor lamp, Bottiglia
Bronze, marble, blown glass
100w halogen bulb
h. 245cm (96½in) d. 42cm (16½in)
Design Gallery Milano, Italy

Roberto Lazzeroni
Floor lamp, Taller Ghost
White porcelain, metal
150w halogen bulb
h. 195cm (77⅜in) w. 36cm (14⅛in) d. 24cm (9½in)
Luminara, Italy

Alfredo Häberli
Floor lamp, Carrara
Polyester resin
Fluorescent / metal halide bulb
Luceplan, Italy

Large
h. 210cm (82½in) di. 20cm (7⅞in)

Small
h. 180cm (70⅞in) di. 20cm (7⅞in)

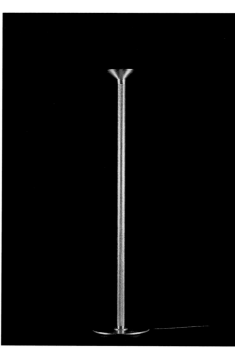

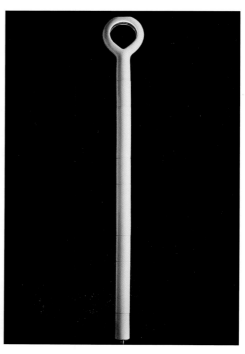
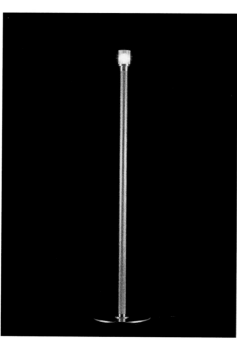

Matthias Bader
Floor lamp, Matteo
Matt aluminium
12w bulb
h. 150–200 cm (59–78³⁄₄in) d. 25cm (9⁷⁄₈in)
Pallucco Italia SpA, Italy
Prototype

Marco Carenini
Lamp, Cap
Aluminium
Globe Energy Saver 21w bulb
h. 160cm (63in)
Prototype

King Miranda Associati
Combined reading/indirect lamp, Diogenes
Aluminium
50w/150w halogen bulb
h. 176.6cm (69½in)
Belux AG, Italy

Diogenes is a novel idea from King Miranda Associati. It is a combined reading/indirect lamp. A high performance halogen bulb serves for general lighting. Integrated into the body of the lamp is an additional dazzle-free and directible 50 watt low-voltage halogen lamp for direct reading light. Both light sources can be controlled independently of one another.

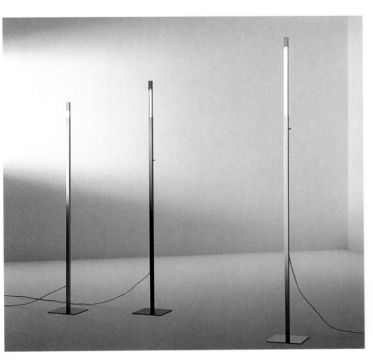

Tobias Grau Design
Table lamp, Bill
Aluminium
50w halogen bulb
h. 52cm (20½in) l. 69cm (27⅛in)
Tobias Grau, Germany

Tobias Grau Design
Table lamp, Soon
Polycarbonate
50w halogen bulb
h. 52cm (20½in) l. 67cm (26⅜in)
Tobias Grau, Germany

Andreas Ostwald and Klaus Nolting
Floor lamp, Minyas
Chromium plated steel tubing, opaline glass
100w halogen bulb
h. 158cm (62⅛in) w. 97cm (38⅛in) di. 24cm (9⅜in)
ClassiCon, Germany

Joan Gaspar Ruiz
Task lamp, Atila
Aluminium, polycarbonate
100w incandescent, 13w fluorescent
h. 90.5cm (35⅝in)
Marset Iluminacion, Spain

Yaacov Kaufman
Table and floor lamp, Naomi
Aluminium and steel
60w incandescent/75w halogen bulb
h. 141cm (55½in) (large)
Lumina Italia, Italy

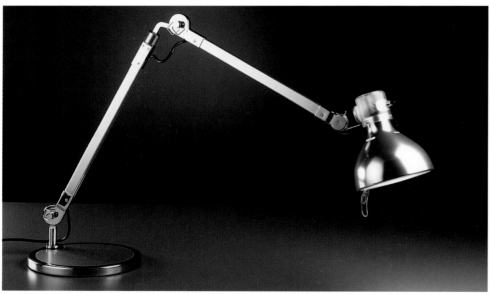

Zumtobel
Wall/floor light, Ledos
Stainless steel, matt glass
24v
l. 10cm (3⁵/₈in) w. 10cm (3⁵/₈in) d. 4cm (1¹/₂in)
Zumtobel Staff AG, Austria

Zumtobel
Spotlight, Phaos
Die cast aluminium
64 SMD-LEDs
l. 20cm (7¹/₂in) w. 20cm (7¹/₂in) d. 3cm (1¹/₈in)
Zumtobel Staff AG, Austria

Zumtobel
Ceiling light, Active Light Field
Metal, glass
Fluorescent bulbs
l. 37–46cm (14–17³/₈in) w. 37–64.75cm (14–24¹/₂in)
Zumtobel Staff AG, Austria

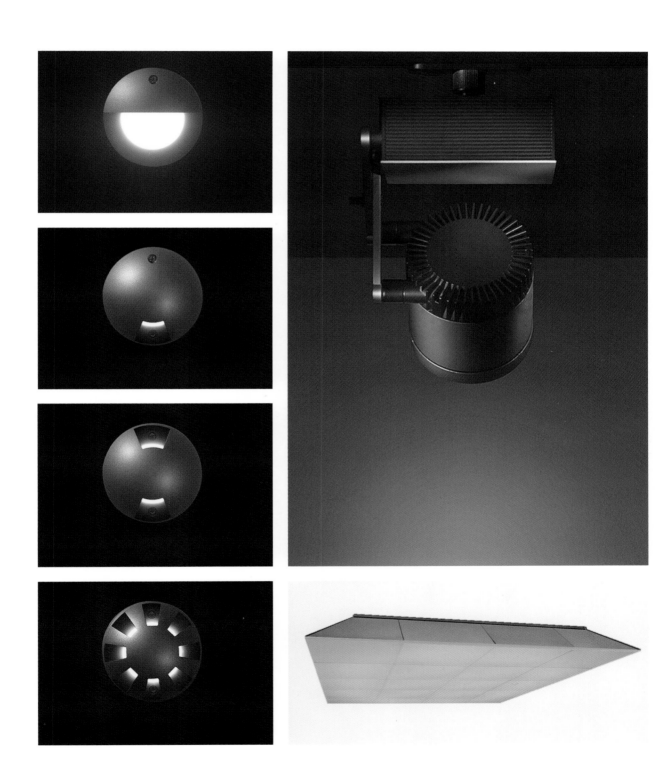

Tableware

It is notable how little the tableware selections of *The International Design Yearbook* have varied over the years. Personal taste has obviously led to some changes in emphases – sometimes on decoration, sometimes the functional austerity of Scandinavian design. However, in the end, a glass is always a glass and a fork a fork. Mario Bellini (Guest Editor of the *Yearbook* in 1990 and 1991) has remarked that any item which is closely related to the human body, or for that matter rituals of society, will always remain similar to those that have preceded it. Even though our lifestyles have changed through time – MIT research indicates that in the 1950s the average family spent one hour around the table at an evening meal while today it is more like 16 minutes – this doesn't seem to have affected the style of flatware.

More than any other category, tableware has long been associated with the craft movement. As such, it is more subject to artistic influences than design. In this discipline, function and content are almost identical and there are few new techniques developing. The surface qualities of objects are what give them life and distinguish one from another.

A piece of tableware has often grown from years of tradition and has developed its form simply because that form does the best job. Claret glasses or brandy balloons are the shape they are for a reason – the wine or liqueur has the chance to breathe and mature in them. Jean Nouvel has written 'Nothing fills me with more horror than seeing someone try to change the design of a wine glass ... to fit a new and more fashionable style ... modifying them for commercial, marketing or economic reasons is a dilution of their content, a banalization of design' (Introduction, *The International Design Yearbook 1995*).

Michele De Lucchi gives as much importance to the design of a vase as to a piece of furniture or an electronic product. Writing in *Domus* (Issue 819), in October 1999, he stated 'I think it is very important to have a mixed balance of things. Because in that way it remains equally positive in my mind to design a vase for flowers or a multipurpose office machine.' His attitude is reflected in his considered selection for this section of the book.

De Lucchi enjoyed both traditional manufacture as in the Arnolfo di Cambio series as well as the beautifully crafted hand-blown glass pieces of Aldo Cibic. Use of experimental material is also present in his choice. Oz Design's Lovenet centrepiece uses rayon fibre hardened by epoxy resin while straw bowls are an interesting departure for Kristiina Lassus. Innovative and practical designs are represented by Massimo Lunardon's party glass and Thomas Rosenthal's 'Short Break on the Run' picnic set. Recognizing the importance of craft in tableware, De Lucchi wanted to feature the traditional Indian lost wax technique of Satyendra Pakhalé and the one-off pieces of hand-blown Murano glass by Emmanuel Babled. The design, however, which both wowed Milan and stands out in the tableware selection of the *Yearbook* was Ron Arad's 'Not Made by Hand, Not Made in China'. This represents the only technological advance in this section. By using computer-controlled laser beams (see p141), Arad and his team have literally 'grown' a range of tableware and luminaires in a tank.

The future of tableware, with the exception of a few changes in fashion and the use of new materials, is likely to continue much as usual. However, there is a new star on the horizon – 'smart' flatware. MIT's Media Lab 'Counter Intelligence Project' is investigating domestic technologies and has patented a plate that can actually calculate calories.

Jean-Marie Massaud
Pepper mill, Pepe
Polypropylene, ceramic
h. 17.5cm (7in) di. 7.5cm (3in)
Authentics, Germany

Marc Newson
Salt and pepper mills, Gemini
Beech wood
h. 12.5cm (4⁷⁄₈in) di. 8.5cm (3³⁄₈in)
Alessi SpA, Italy

Ka-chi Lo
Toothpick holder, Polar Molar
Ceramic
h. 9cm (3½in) w. 8cm (3⅛in)
d. 3cm (1⅛in)
Prototype

Sebastian Bergne
Cutlery, Slope
Stainless steel
Driade, Italy

Paola Navone
Soup ladle and serving spoon, Paloma
Stainless steel
h. 33.6cm (13¼in) h. 22.7cm (9in)
Driade, Italy

Borek Sipek
Cutlery, Hebe
Stainless steel, plastic
Arzenal, Czech Republic

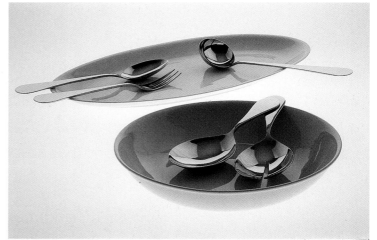

Claus Jensen
Spoons and spatulas, Eva Solo
Aluminium and plastic
h. 30cm (11⅞in) di. 9cm (3½in)
Eva Denmark A/S

Enzo Mari
Bowls, Campanella
Barium crystal
Various sizes
Arnolfi di Cambio, Italy
Limited batch production

Marta Laudani, Marco Romanelli
Tableware, Mediterraneo
Gres
Various sizes
Driade SpA, Italy

Tableware

Gijs Bakker
Tray, Balance
Steel
h. 3cm (1¹⁄₈in) di. 41cm (16¹⁄₈in)
Meccano, The Netherlands

**Bastiaan Arler
Tray, Teevee
Ceramic
h. 3.5cm (1³/₈in) l. 40cm (15³/₄in) w. 30cm (11⁷/₈in)
Britefuture, Italy
Limited batch production**

Tableware

Britefuture is a young company based in Italy. Its philosophy is to 'design and commercialize small objects using semi-industrial techniques for a refreshing approach to design for the home'. Its 'Teevee' plate/tray is a humorous yet practical solution for the 'couch potato'.

Thomas Rosenthal
Picnic set, Short Break on the Run
Plastics
h. 35cm (13⅝in) l. 36cm (14⅛in)
Rosenthal AG, Germany

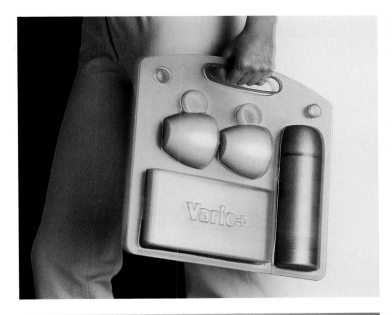
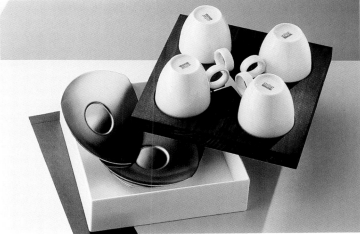

Ely Rozenberg
Centrepiece, Tamnun
Harmonious steel, zipper
d. 7cm (2¾in) di. 58.5cm (23in)
Oz, Italy
Limited batch production

Ole Palsby
Basket, Concept
Stainless steel wire
WMF, Germany

Alessandro Bianchini
Centrepiece/container, Lovenet
Rayon fibre, epoxy resin
h. 48cm (18⁷/₈in) w. 65cm (25¹/₂in) l. 57cm (22¹/₂in)
O2, Italy
Limited batch production

More art-work than tableware, the organic forms of Alessandro Bianchini's containers hang from the ceiling like giant wasps' nests. Each piece is unique, the pliable copper core of the fibre is spun and woven in a series of random patterns which are then locked, by the addition of epoxy resin as it hardens, into the final shape.

Tableware

Andrea Branzi
Fruit basket, Solferino
Stainless steel
h. 23cm (9in) di. 23cm (9in)
Alessi SpA, Italy

Kristiina Lassus
Centrepiece, Strawbowls
Straw
Large: h. 13.5cm (5¹/₈in) di. 30cm (11⁷/₈in)
Small: h. 10cm (7¹/₂in) di. 22cm (8⁵/₈in)
Alessi SpA, Italy

Andrea Branzi
Small vases, Genetic Tales
Porcelain
h. 18cm (7in) di. 5cm (2in)
Alessi SpA, Italy
Limited batch production

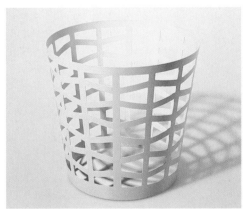

Barbora Skorpilova, Jan Nedved
Vase, Malá zubatá
Duralaluminium
h. 30cm (11⁷⁄₈in) w. 9.5cm (3³⁄₄in) d. 9.5cm (3³⁄₄in)
Giga, Czech Republic
Limited batch production

Henryk Lula
Vases
Terracotta
h. 32cm (13in) di. 46cm (18in)
Limited batch production

Barbora Skorpilova, Jan Nedved
Vase, Rastr
Duralaluminium
h. 30cm (11⁷⁄₈in) w. 9.5cm (3³⁄₄in) d. 9.5cm (3³⁄₄in)
Giga, Czech Republic
Limited batch production

Lemongras
Bowl
Coiled nylon belt
di. 20–50cm (7⁷⁄₈–18⁵⁄₈in)
Lemongras Products, Germany

Ron Arad
Vase/Light, Not Made by Hand, Not Made in China
Polyamide
h. 5–34cm (2–13³/₈in) di.13cm (5¹/₈in)
Limited batch production

'Not Made by Hand, Not Made in China' takes organic design to the extreme. Ron Arad not only gives this range of vases and luminaires 'natural' shapes but nurtures them as they gradually develop in their artificial womb. These pieces are no longer moulded, formed, assembled or honed, because Arad, along with Geoff Crowther, Yuki Tano and Elliott Howes, has pioneered a way of 'growing' designs in a tank by using computer-controlled laser beams.

A three-dimensional model is generated on a computer and it is animated until the desired form has been produced. The model is then sent to a machine which slices it into thousands of horizontal layers. A 'platform' is gradually lowered each time by a depth equal to that of the layers. At each stage the tank which contains the platform is topped-up with the chosen material – either powder or resin, never both. A laser then passes over the new layer of material creating and joining one slice to the previous ones. Over a period of hours, or sometimes days, the process is completed and the item can be removed from its incubator to start life.

Ron Arad Associates has bridged the gap between what can be created and modelled on the computer and what can be finally realized. However, with the rapid progression of technology, it also recognizes that its work in this field is still in its infancy.

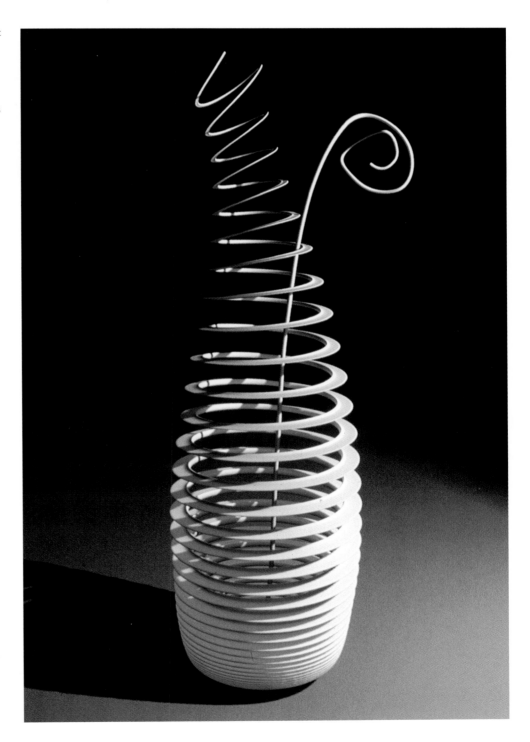

Marc Boase
Bowl, Noodle Bowl
Porcelain
h. 10cm (4in) di. 20cm (7⅞in)
Maplestead Pottery Services, UK

Marc Boase
Bowl, Dip Bowl
Porcelain
h. 10cm (4in) di. 20cm (7⅞in)
Maplestead Pottery Services, UK

Borek Sipek
Dishes, Verna
Porcelain
di. 26cm (10⅛in), 24cm (9½in), 19cm (7½in), 16cm (6⅜in)
Arzenal, Czech Republic

Marco Susani and Mario Trimarchi
Pewter objects, Metallia
Pewter
Average height: 15cm (6in)
Serafino Zani srl, Italy

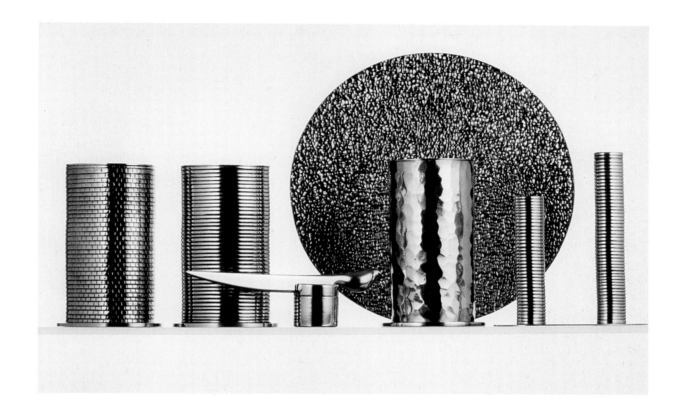

Monica Guggisberg and Philip Baldwin
Glassware, Fili d'Arianna
Blown glass
Large: h. 39cm (15in) di. 19cm (7½in)
Small: h. 21cm (8¼in) di. 28cm (11in)
Venini, Italy
Limited batch production

Nucleo
Candle, Cerapura
Paraffin
h. 41cm (16⅛in) di. 21cm (8¼in)
Nucleo Global Design Factory, Italy

For the last couple of years we have seen countless reinterpretations of the candle-as-candlestick idea. In 2000, Droog Design added life to the concept with its 'Paraffin Table' by Timo Breumelhof (not illustrated) which burns away as it is used. Nucleo Global Design Factory has also succeeded in reworking the concept; by adding colour and texture to the wax and shaping the candle in the form of a vase it has conceived a design, 'Cerapura', which engages the senses. The light of the flame transforms the object into a luminous centrepiece.

Ronan Bouroullec
Candle, Candle Syst
Paraffin
h. 45cm (17⅝in) w. 25cm (10in) l. 35cm (13⅝in)
Backstage, France

Jacob de Baan
Candle lamp, Luccichio
Aluminium
h. 22.5cm (8⅞in) di. 20cm (7⅞in)
D4 Industrial Design, The Netherlands

Takashi Ifuji
Candle lamp, Aqmara
Ceramic
h. 13cm (5⅛in) w. 18cm (7⅛in) d. 27.5cm (10⅞in)
Prototype

Takashi Ifuji creates poetry with this candle lamp, 'Aqmara'. As the candle burns away the crescent reflected on the wall waxes to a full moon.

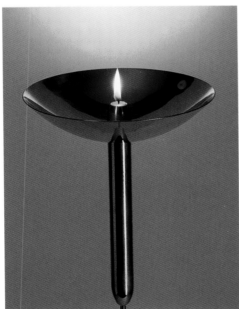

'Luccichio' is one of the designs from the (Non)electrical collection by Jacob de Baan. He has taken the primal form of light source, the flame, as a starting point and designed modern reflector technology around it. Luccichio is a candle holder – a mobile lamp that can be held like a torch, hung from the wall or clamped to any surface.

Massimo Lunardon
Vase, Tulipano
Borosilicate glass
h. 40cm (15¾in) di. 15cm (6in)
Massimo Lunardon & Co, Italy

Massimo Lunardon
Glass, Glass for Party
Borosilicate glass
h. 18cm (7in) di. 5.5cm (2⅛in)
Massimo Lunardon & Co, Italy

Massimo Lunardon has solved those two awful party problems: having a glass in one hand and a plate of food in another and wanting to eat, or, worse still, losing your glass altogether. By suspending the glass around the neck and changing its shape he has simultaneously created an item of jewellery and a new type of drinking vessel. It is important, however, not to get involved in too many bear hugs!

Enzo Mari
Goblet, Campanella
Barium crystal
Arnolfo di Cambio, Italy
Limited batch production

Enzo Mari
Bowls, Campanella
Barium crystal
Arnolfo di Cambio, Italy
Limited batch production

Enzo Mari
Goblet, Stromboli
Barium crystal
Arnolfo di Cambio, Italy
Limited batch production

Enzo Mari
Pitcher, Brocca
Barium crystal
h.22cm (8⁵⁄₈in) di. 6.6cm (2⁵⁄₈in)
Arnolfo di Cambio, Italy
Limited batch production

The lead-crystal manufacturing company of Arnolfo di Cambio was established in 1963. Its design philosophy is to fuse advanced industrial concepts with traditional artisan craftsmanship. Its strong entrepreneurial focus and cultural identity means that it is now established as an industry leader and collaborates with the most prominent figures from the design world. The second edition of the 'Clearline' collection, part of which is published here, illustrates the new artistic direction which the company hopes to take over the next few years. The concept was born out of its association with Prof. François Burkhardt, editor of *Domus* magazine, who wanted to produce a series of glassware where 'essentials were joined with a rigorous aesthetic gain, without losing itself in a minimalism which I believe to be out of place in the art of tableware'.

Borek Sipek
Red wine and white wine glass, Campanula, Magnolia
Calium crystal
Red wine: h. 29cm (11½in) di. 8cm (3⅛in)
White wine: h. 31cm (12¼in) di. 7cm (2¾in)
Arzenal, Czech Republic

Boda Horak
Glass, Stela
Copper, crystal
h. 22cm (8⅝in) di. 10cm (4in)
Anthologie Quartett, Germany
Limited batch production

Boda Horak
Glass, Tekla
Copper, crystal
h. 22cm (8⅝in) di. 10cm (4in)
Byra Interior Objects, Germany
Limited batch production

Tableware

Aldo Cibic
Chalice
Glass
h. 27cm (10½in), 22.5cm (8⅞in), 21cm (8¼in), 20cm (7⅞in)
Cibic & Partners, Italy

Aldo Cibic
Candle holder
Glass
Large: h. 40.5cm (16in)
Medium: h. 34.5cm (13½in)
Small: h. 27.5cm (10⅞in)
Cibic & Partners, Italy

Aldo Cibic
Pitcher
Glass
h. 34.5cm (13½in)
Cibic & Partners, Italy

Aldo Cibic
Bottle
Glass
h. 42cm (16½in)
Cibic & Partners, Italy

Christian Ghion
Vase, Insideout
Hand-blown glass
h. 40cm (15³/₄in) di. 30cm (11⁷/₈in)
XO, France

Vanessa Mitrani
Vase, La Mome Catch-catch
Glass, metal
h. 27cm (10⁵/₈in) di. 25cm (9⁷/₈in)
Ligne Roset, France

Vanessa Mitrani
Vase, Lola Molotov
Glass, metal
h. 24cm (9½in) di. 14cm (5½in)
Ligne Roset, France

Karim Rashid
Decanter, Blob
Glass
h. 26cm (10¼in)
di. 18cm (7⅛in)
Leonardo, Germany

Karim Rashid
Decanter, Spoo
Glass
h. 25cm (9⅞in)
di. 14.5cm (3⅝in)
Leonardo, Germany

Karim Rashid
Decanter, Dive
Glass
h. 31cm (12¼in)
di. 13cm (5⅛in)
Leonardo, Germany

Karim Rashid
Bowl, Flip
Glass
h. 25cm (9⅞in)
di. 15cm (6in)
Leonardo, Germany

Karim Rashid
Bowl, Loop low
Glass
h. 20cm (7⅞in)
di. 29cm (11⅜in)
Leonardo, Germany

Karim Rashid
Bowl, Loop high
Glass
h. 25cm (9⅞in)
di. 23.5cm (9¼in)
Leonardo, Germany

Since the formula of mixing salt (sodium), bones (calcium) and sand (silica) to make glass was discovered over 4,500 years ago it has become a lasting source of inspiration for artisans and designers. It is an insulator, has an elastic quality, is transparent and – although it looks light – cubic metre for cubic metre, its weight is equivalent to that of Portland stone. Glass has a life and, for a group of designers, has become the object itself.
Christian Ghion, Karim Rashid, Vanessa Mitrani and Danny Lane have all produced important pieces this year. Lane's unmistakable and powerful designs play on the paradoxical effect of the material – it weighs approximately 2570kg per cubic metre yet it has no visual mass. He has developed a technique of achieving the optimum refraction of light. By using a new low iron glass the spectral quality of his work is enhanced; he says, 'now it means I can make solid volumes on a large scale and the prismatic events in these techniques are much stronger'.
　Karim Rashid believes glass to be the most beautiful natural material in the contemporary environment. The sensual curves and ethereal quality of the graded colours of his Leonardo series are intended to 'emphasize the need for beauty and material, through soft, solid, ergonomic, sensual, desirable forms'.

Emmanuel Babled
Glassware, Primaire N° IX
Hand-blown glass
Average di. 45cm (17¾in)
Covo srl, Italy
Limited batch production

Emmanuel Babled
Glassware, Edizioni
Hand-blown glass
h. 19–40cm (7–15¾in) di. 50cm (19⅝in)
Covo srl, Italy
Limited batch production

Danny Lane
Bowl, Crab Bowl
Kiln formed glass
h. 18cm (7in) l. 72cm (28½in) d. 40cm (15¾in)
One-off

Perhaps Emmanuel Babled has the greatest respect for the character, poetry and mysteries of glass. His hand-blown vases are all produced in Murano, Italy, by a Master Craftsman, to Babled's design. However, he believes that it is the glass itself which determines the outcome of a piece, the 'Maestro' and designer being merely servants to the wishes of the material. A large part of the end result cannot be predetermined and it is this tension between the human desire and the moment of creation which he considers to be the most interesting aspect of working with glass. Changes can only be made at the last moment, in the furnace, while the glass is still malleable; the next second it crystallizes and becomes 'fragile for the rest of eternity' (EB). In his designs, he uses this instant to 'capture a fragment of emotion', and he believes in not only producing a beautiful object with an original and modern form but in giving that object a personal significance. He does not follow style or fashion but expresses a different sensation with every vase he produces.

Ettore Sottsass
Centrepiece, Namus
Lead crystal, black marble
h. 18.9cm (7½in) di. 21cm (8¼in)
Arnolfo di Cambio Comp. Italiana del Cristallo srl, Italy
Limited batch production

Ettore Sottsass
Vase, Manaus
Lead crystal
h. 34.8cm (13¾in) di. 13cm (5⅛in)
Arnolfo di Cambio Comp. Italiana del Cristallo srl, Italy
Limited batch production

Oscar Tusquets Blanca
Hors d'oeuvre dish container, Tableta
Lead crystal
h. 3.4cm (1⅜in) w. 29.8cm (11¾in) l. 29.8cm (11¾in)
Arnolfo di Cambio Comp. Italiana del Cristallo srl, Italy
Limited batch production

Oscar Tusquets Blanca
Candle holder with flower stand, Fioreluz
Lead crystal
h. 34cm (13⅜in) w. 28cm (11in)
Arnolfo di Cambio Comp. Italiana del Cristallo srl, Italy
Limited batch production

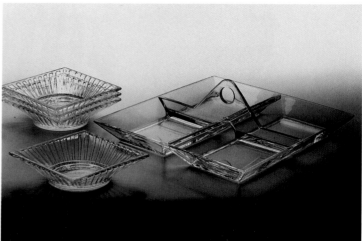

Paolo Zani
Flower vase, Sumo
Pyrex and hand-blown glass
h. 40cm (15¾in) di. 13cm (5⅛in)
FontanaArte, Italy

Depending on the length of the flower stem, the water container of Paolo Zani's flower vase 'Sumo' can be removed, the external cylinder inverted, and the container returned to a lower position.

Carsten Joergensen
Glass dish, Hot Pot
Borosilicate glass
di. 22.1cm (8¾in)
Bodum, Switzerland

Tableware

Satyendra Pakhalé
Table top object, C:da
Recycled metal, wax
h. 15.5cm (6⅛in) di. 10cm (4in)
Limited batch production

Satyendra Pakhalé
Table top object, K:dai
Recycled metal, wax
h. 5.5cm (2⅛in) di. 18cm (7⅛in)
Limited batch production

Michele De Lucchi's interest in the combination of design and craft resulted in him selecting Satyendra Pakhalé's collection of centrepieces and vases. Pakhalé has refined an ancient Indian metal casting technique to produce contemporary objects. A core is made with a mixture of fine clean sand and clay. Traditionally, goat's dung is then soaked in water and ground and mixed with clay in equal proportions. This soft mixture forms the base mould. Once dried, it is used to create a wax pattern. A special natural wax is melted over an open fire and strained through a fine cloth into a basin of cold water where it becomes solid. Care has to be taken to keep the wax absolutely clean and free of impurities. Next, it is squeezed through a sieve and recovered in the form of wax 'wires' (thick or thin as desired). Each of these wires is wrapped around the core, one after another, until the whole surface is covered. Originally, the artisan would have sat in the sun to let the clay core and the wax coating warm up uniformly. The whole form is finally covered in a mixture of equal parts clay, sand and cow dung and fired. Base metals – brass, bronze, copper etc. – are melted together and poured into the fired clay mould to form the metal objects, the wax being lost in the process.

Textiles

Of all the chapters in *The International Design Yearbook*, this is the one which is universally forbidding for the Guest Editor to select. Not only is the subject often one with which he or she is unfamiliar but it is also the most difficult to judge without having the item physically present. Textiles are tangible – a two-dimensional version does not allow examination of texture, which is fundamental to the assessment of a fabric. The decision, therefore, ultimately has to be made on the grounds of aesthetic considerations and innovation.

This year the choice was broad: mass-manufactured upholstery by Bute and De Padova; computer-generated designs from Carol Westfall; creative wall-hangings by Yehudit Katz; Masayo Ave's and Santos and Adolfsdóttir's experimentation in the use of materials; the creative designs of Claudy Jongstra and the wide selection of rugs from Asplund and Ligne Roset to mention just a few. For once Japanese entries did not overwhelm the other nominations and apart from the work of the Nuno Corporation, which combines traditional native craft with the latest in materials and production techniques, are not present in the pages which follow.

The most interesting trend which seems apparent from the selection is the movement away from the traditional definition of fabric. Boundaries have become disparate – we have textile as furniture (Sodeau's Red Rug and Gavoille's 'Kloc'), textile as product (CP's 'City Tent') and, most significantly, 'smart' textiles.

As was noted in last year's *Yearbook*, advances in materials and technology are becoming progressively important. Textiles are moving into the 21st century. Not only do we have yarns and fabrics created from industrial materials but the arrival of 'smart' ware has blurred the parameters between textile and product. One can now buy an 'Antistress Car Coat', the lining of which moulds itself to the contours of the person wearing it to support the neck and the small of the back, or a 'Temperature Jacket' with NASA-improved lining which monitors body temperature, absorbing or releasing heat to keep the body at a steady level. Levi's is even in the process of developing a denim jacket with a built-in keyboard and synthesizer.

Pharmaceutical textiles have been developed: anti-static fabrics for patients with pacemakers or moisturizing materials to be worn by burns victims. The Florence-based Lineapiu Group has produced glow-in-the-dark rubber yarn which stores light and releases it, as well as a carbon-fibre yarn called 'Relax' which apparently reduces stress by protecting the body from some of the electromagnetic waves issuing from household appliances.

The selection also features the tablecloth of the future. A concept design developed by Philips in its 'Culinary Art – Searching for Total Culinary Enjoyment' research programme – which applies new digital technology to the traditional meal time – this linen cloth is fully washable yet includes an integrated power circuit which supplies electricity to all appliances placed on it.

Where could 'smart' textile development lead us? Futuremode, a trend-forecasting firm in New York, may have the answer, having recently published findings of concept work in Japan which includes the development of a metallic mesh fabric that can be folded, written on, backlit or infused with information.

Marc Newson
Carpet, Marc
100% New Zealand wool, handtufted
Asplund, Sweden

Marc Newson
Carpet, Marc De Luxe
100% New Zealand wool, handtufted
Asplund, Sweden

Maria Kaaris
Carpet, Soap Bubbles
100% New Zealand wool, handtufted
l. 160cm (63in) w. 210cm (82⁵/₈in)
Asplund, Sweden

Alfredo Häberli
Carpet, Carpet Lines
100% New Zealand wool, handtufted
Asplund, Sweden

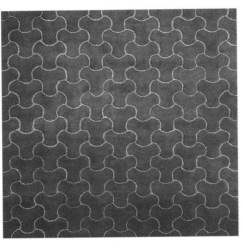
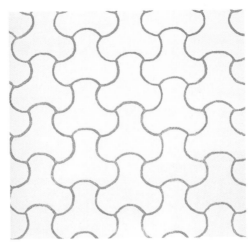

Alfredo Häberli
Carpet, Lines
100% New Zealand wool, handtufted
Asplund, Sweden

Anki Gneib
Carpet, Cheese
100% New Zealand wool, handtufted
Asplund, Sweden

Lars Bergström, Mats Bigert
Carpet, Egg
100% New Zealand wool, handtufted
Asplund, Sweden

**Masayo Ave
Carpet, Silent
Industrial wool felt
w. 80cm (31½in) l. 160–180cm (63–70⅞in)
Atrox GmbH, Switzerland
One-off**

**Masayo Ave
Wall carpet, Quadretti
Industrial wool felt
Atrox GmbH, Switzerland
Prototype**

Masayo Ave's self assembly rugs harmonize tradition and innovation – the long-practised process of weaving has been given a new twist by using woollen industrial felt. Ave has been making an in-depth study of material since 1995. She searches among substances normally associated with industrial manufacture and liberates their aesthetic qualities. Currently she is interested in polyester material (which can be produced CFC-free as well as recyclable), un-woven textiles such as felt and filter material, artificial marble, PET and nylon. By using the intrinsic potential of these elements, Ave creates individual pieces which she says are 'often unexpected and joyful discoveries'. Anxious never to impose a design onto a material, she 'waits for it to speak to her' and as such her objects often engage directly with the viewer; they not only function but appeal to the emotions.

Masayo Ave
Cushion, Cool
Polyester foam, tulle
Large: w. 42cm (16½in) l. 42cm (16½in)
Small: w. 32cm (12½in) l. 32cm (12½in)
Atrox GmbH, Switzerland
Prototype

Claudy Jongstra
Textile
Merino wool, silk metal organza
Not tom dick & harry, The Netherlands
Prototype

Claudy Jongstra
Textile
Wool, merino, karakul, raw silk
Not tom dick & harry, The Netherlands
Prototype

Claudy Jongstra
Textile
Merino wool, raw linen, silk organza, karakul
Not tom dick & harry, The Netherlands
Prototype

Claudy Jongstra
Textile
Merino wool, tulle
Not tom dick & harry, The Netherlands
Prototype

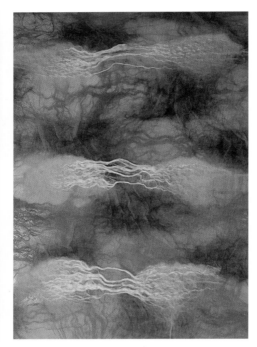
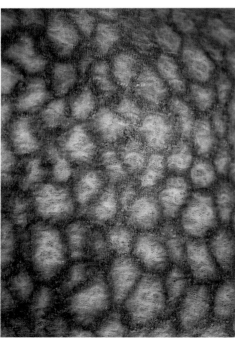

Claudy Jongstra of Not tom dick & harry works entirely with natural material, primarily felt, to which she adds raw silk, linen, camel, cashmere and wool. The fabrics she creates are 'Beauty and the Beast' – they have the primeval look of the caveman touched with the delicacy and beauty of a gossamer wing. She has a repertoire of over 500 designs. More sophisticated, smoother fabrics might be a combination of alpaca, merino and silk metallic organza, while hairier, wilder ones might be a mix of merino and raw linen. Jongstra says, 'Basically, felt is the oldest textile in the world and thus does not have a modern image. I wanted to find out more about how to develop some contemporary versions. I want to respect its original character, its strength, so that is why I work with raw, untreated materials.' To control the quality, colour and mix of wool she uses, Jongstra owns her own herd of rare species sheep, and any industrial process is undertaken on machines which she has had specially made. Clients include international fashion designers Donna Karan and John Galliano; industrial designer, Hella Jongerius and American architect Will Bruder. She has recently supplied the fabrics for the Jedi warriors in the latest 'Star Wars' film.

Norma Starszakowna
Textile, Acts of Beauty A
Heat reactive and print media
h. 350cm (137³/₄in) w. 50cm (19⁵/₈in)
Prototype

Norma Starszakowną
Textile, Acts of Beauty B
Heat reactive and print media in silk organza
h. 350cm (137³/₄in) w. 50cm (19⁵/₈in)
Prototype
Limited batch production

Emmanuele Ricci
Furnishing fabric, Wire
51% cotton, 49% acrylic
w. 140cm (55⅛in)
Lorenzo Rubelli SpA, Italy

Emmanuele Ricci
Furnishing fabric, Rise
51% viscose, 25% polyester, 24% silk
w. 130cm (51⅛in)
Lorenzo Rubelli SpA, Italy

Emmanuele Ricci
Furnishing fabric, Bambusa
51% cotton, 49% viscose
w. 140cm (55⅛in)
Lorenzo Rubelli SpA, Italy

Emmanuele Ricci
Furnishing fabric, Helix
51% viscose, 25% polyester, 24% silk
w. 130cm (51⅛in)
Lorenzo Rubelli SpA, Italy

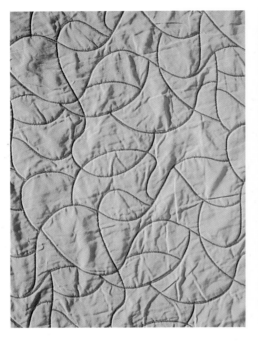

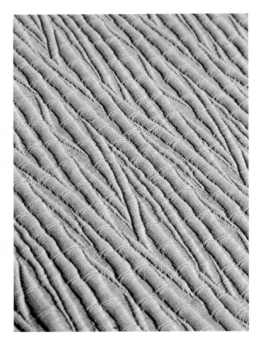

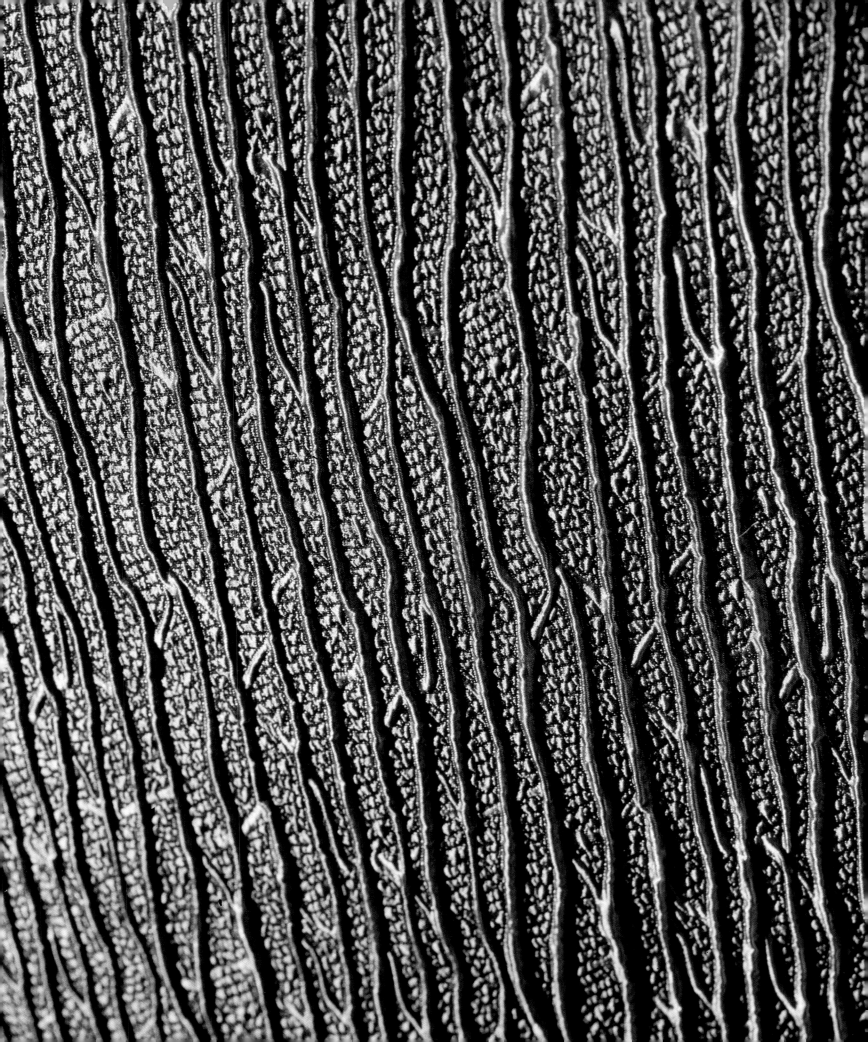

Ane Lykke
Carpet/rug, Zen Se
Woven paper
h. 3.5cm (1³⁄₈in) w. 120cm (47¼in) l. 200cm (78¾in)
Prototype

Javier Mariscal
Rug
100% pure New Zealand wool
l. 170cm (67in) w. 240cm (94½in)
Desso, The Netherlands

Jeffrey Bernett
Carpet, Man
100% wool, handtufted
l. 170cm (67in) w. 240cm (94½in)
Ligne Roset, France

Pascal Mourgue
Carpet, Smala
100% wool, handtufted
l. 200cm (78¾in) w. 200cm (78¾in)
Ligne Roset, France

Jean-Charles de Castelbajac
Carpet, On Pax Joy
100% wool, handtufted with carved design
l. 170cm (67in) w. 240cm (94½in)
Ligne Roset, France

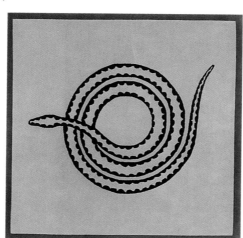

Michael Sodeau
Rugs, Red Rug, Blue Rug
Wool
w. 100cm (39³⁄₈in) l. varies
Christopher Farr, UK

Kristian Gavoille
Carpet, Kloc
100% wool, handtufted
l. 200cm (78³⁄₄in) w. 200cm (78³⁄₄in)
Ligne Roset, France

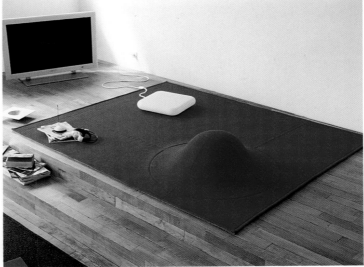

Stefano Marzano, Philips Design
Tablecloth, Interactive Tablecloth
Fabric, electrical circuits
Philips Electronics BV, The Netherlands
One-off

The interactive tablecloth is one of the items produced by Philips in its 'Culinary Art' series. The integrated power circuit is woven into the washable linen tablecloth and supplies electricity to all appliances placed on the table. The surface of the cloth is cool at all times but it is able to keep specially adapted ceramic plates warm.

CP Company
City Tent/Mac
Weather-proof nylon mesh
SPW, Italy
Limited batch production

The Spring/Summer 2000 collection of CP Company is not content to be either just fashion or just textile. Its 'Transformables: beyond clothing, an ironic aspiration to freedom' is a series of outfits which 'transform' into something quite different – a long cloak turns into a kite with a thread to attach to the body and a jacket changes into a multi-pocketed rucksack. The cape featured here comes with a small bag containing lightweight aluminium poles which fold up into it. These are then inserted into the seams of the cape and a 'city tent' (this product is not gear-tested for strength and extreme weather conditions) is formed.

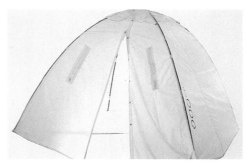

Jasper Morrison
Fabric, Melrose
Wool
Bute Fabrics Ltd, Scotland, UK

Jasper Morrison
Fabric, Tiree
Wool
Bute Fabrics Ltd, Scotland, UK

De Padova
Lino 2000
54% linen, 40% cotton, 6% polyamide
De Padova, Italy

De Padova
Upholstery fabric, Melange 2000
80% cotton, 10% viscose, 6% nylon, 4% linen
De Padova, Italy

Heinz Röntgen
Decorating fabric, Bijoux
50% polyester, 50% pes-metal
w. 170cm (67in)
Nya Nordiska Textiles GmbH, Germany

Heinz Röntgen
Decorating fabric, Fukaso
33% Linen, 33% polyamide 34% polyester
w. 152cm (59¾in)
Nya Nordiska Textiles GmbH, Germany

Yehudit Katz
Textile, Ikat
Waxed linen, cotton, bamboo sticks
h. 226cm (89in) w. 75cm (29½in)
One-off

Yehudit Katz
Textile, Degradèe
Linen, copper threads
h. 75cm (29½in) w. 68cm (26¾in)
One-off

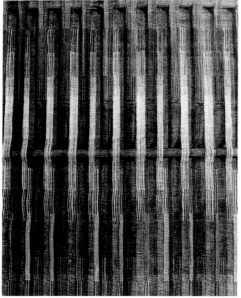
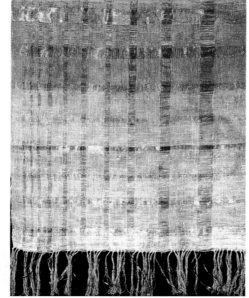

Leo Santos-Shaw, Margaret Adolfsdóttir
Textile, SA08B2
Paper, nylon
w.150cm (59in)
One-off

Leo Santos-Shaw, Margaret Adolfsdóttir
Textile, SA0012
Polyester, polyamide
h. 400cm (157½in) w. 100cm (39⅜in)
One-off

Tuttu Sillanpää and Elina Huotari
Carpet, Polku I, Polku II
Pure wool felt
l. 150cm (59in) w. 220cm (86⅝in) di. 160cm (63in)
Verso Design, Finland

Thomas Sandell
Carpet, Air
100% New Zealand wool, handtufted
Asplund, Sweden

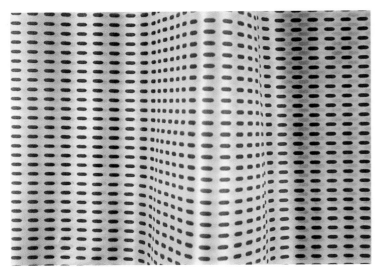
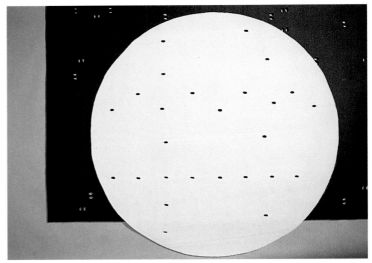
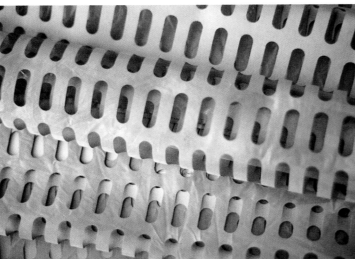

Carol Westfall
Textile, Vegetable-Cucumber Wilt
Digital print on cotton
w. 148.5cm (58½in) l. 103.5cm (40¾in)
One-off

Carol Westfall's designs are generated on computer and digitally printed on cotton.

Carol Westfall
Textile, Animal-Protozoan
Digital print on cotton
w. 148.5cm (58½in) l. 274cm (108in)
One-off

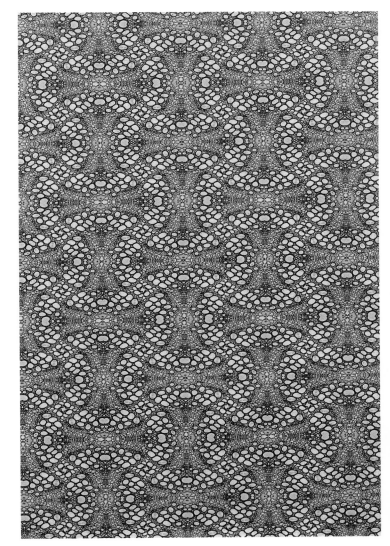

Christine Van der Hurd
Handtufted area rug, Verdi
100% New Zealand wool
w. 304.8cm (120in) l. 213.3cm (84in)
One-off

Christine Van der Hurd
Handtufted area rug, Portals
100% New Zealand wool
w. 152.4cm (60in) l. 213.3cm (84in)
One-off

Reiko Sudo
Fabric, Rubber Band Scatter
Linen
w. 112cm (44in)
Nuno Corporation, Japan

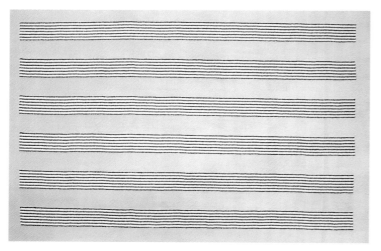

Kazuhiro Ueno
Fabric, Masking Tape
Cotton
w. 92cm (36¼in)
Nuno Corporation, Japan

Reiko Sudo, Ryoko Sugiura
Fabric, Southern Cross
Polyester
w. 112cm (44in)
Nuno Corporation, Japan

Yoko Ando, Reiko Sudo
Fabric, Random House
Polyester
w. 112cm (44in)
Nuno Corporation, Japan

This year Reiko Sudo's range of fabrics for Nuno Corporation reflects the artisan tradition of Japanese textile design. The descriptions of the pieces are more applicable to a creative design class than to mass-produced fabrics in one of the best known textile manufacturers in Japan. 'Rubber band-like shapes are printed on linen in a mixture of acrylic and silicone. If only physics class were this much fun' ('Rubber Band Scatter'); 'strips of masking tape are affixed to the fabric, sprayed over with colour and peeled away leaving hard-edged patterns' ('Masking Tape'); tiny glass beads affixed with industrial glue' ('Southern Cross').

Products

'I have always followed the wind of change, and I still do. Because that seems to me a condition necessary to the creation of architecture and design' (*Domus*, October 1999). From his early career and the foundation of 'Cavart', through his involvement with Memphis, and today in his interest in the prototypical world that the development of new technologies such as Wireless Application Protocol (WAP) and Bluetooth have brought to the arena of electronic products, Michele De Lucchi is consistently found at the cutting edge of major developments. Since his collaboration with Olivetti in the late 1970s, and more recently in his association with Matsushita and Siemens, De Lucchi has recognized the importance of the in-house design team in the development of new technologies. His company adopts an holistic philosophy which not only encapsulates the aesthetics of an item but also puts great emphasis on the development process 'from analysis to concept and from research to realization'. When discussing the merits and drawbacks of manufacturing products at the dawn of the information and telecommunication age, De Lucchi declared that those who will succeed will do so because they have a sound research programme. 'The degree of success today no longer depends on production capacities, but on the capacity to produce ideas, to develop and fulfil them' (*Domus*, October 1999).

As socio-cultural trends change and a fuller, faster and more varied lifestyle emerges, smart products are being developed by all major manufacturers with the aim of liberating consumers' time to pursue more creative activities. Whirlpool is working on an internet fridge with food-tracking capabilities. It can recognize its contents and request deliveries directly from the supermarket via e-mail. Microwaves are being designed which can read barcodes and weigh and cook ready-made meals for the appropriate amount of time. Sunbeam has recently announced a range of networked appliances, 'Thalia' – including coffee-maker, alarm clock, mixer and bathroom scales – which has been designed to interact with us to meet our needs as soon as we awaken. A recent publication, *When Things Start to Think* by Neil Goshenfeld, a professor at the Massachusetts Institute of Technology, even prophesies that it will not be so long before a networked kitchen could provide insurance companies with dietary information to gauge premiums.

As computers threaten to dehumanize our world, ways have to be found to bring them back into line, to find a balance between technological advancement and the human touch. Since Apple brought out the iMAC we have already seen the outward design of products simplify and become more user-friendly while their inner complexities develop.

Similarly, Philips has long been associated with mixing innovative electronics with traditional aesthetics (the Philips-Alessi range was revolutionary in bringing colour and human warmth back into the high-tech sterile kitchen of the 1990s) and it is presently involved in a long-term strategic design project with the aim of humanizing the explosion of internet-based services and mobile communications. Philips Consumer Communications and Philips Design are consulting operators, retailers, media and consumers to define and develop a more human-focused future. They believe that the market wants to retain a sense of control and recognize the need for human contact. Consumers want to control their products, not the other way round. Stefano Marzano, the Managing Director of Philips Design, writes that ideally 'the new intelligence that will surround us will nestle in "assistant" or "companion" objects that "know" us, learning to fit in with our likes and dislikes, rather like servants in former times, leaving their employers free to engage in the arts, scientific research, social intercourse, entertaining visitors, travel and charitable activities.'

The *Yearbook* selection highlights some of these points. Leon@rdo by Ariston with WRAP (Web Ready Appliance Protocol) brings the internet to household appliances. Matsushita has produced a range of products, using traditional and familiar materials, which changes the way we consider 'breakfast' both physiologically and psychologically. Jam, in collaboration with Whirlpool and Corian, has conceived a fully interactive kitchen with soft lines and warm textures and Philips has brought out the Café Duo which it describes as not looking like a machine but more like a friendly waiter standing against the wall holding coffee cups on a tray.

Although the fully automated and internet-linked home of the future has not yet arrived, it is just around the corner. At the time of going to press a German billionaire was advertising for a technologically adept and multi-lingual family to live in a house he has built in Switzerland. Among the many modern WRAP appliances it contains, are a self-ordering fridge and a lavatory that can diagnose potential diseases. The daily activities of the family will be broadcast on the internet (www.futurelife.ch). We can only hope that this experiment will prove an inspiration and not a warning. ('Luxury life under gaze of world', *The Times*, June, 2000).

Marc Berthier, Design Plan Studio
Radio, Pebble Objects
Aluminium jacketing, P.P.
h. 2cm (³/₄in) w. 6cm (2¹/₂in) l. 9cm (3¹/₂in)
Spirix-Lexon, France

Marc Berthier, Design Plan Studio
Voice memo, Pebble Objects
Aluminium jacketing, P.P.
h. 2cm (³/₄in) w. 6cm (2¹/₂in) l. 9cm (3¹/₂in)
Spirix-Lexon, France

Marc Berthier, Design Plan Studio
Calculator, Pebble Objects
Aluminium jacketing, P.P.
h. 2cm (³/₄in) w. 6cm (2¹/₂in) l. 9cm (3¹/₂in)
Spirix-Lexon, France

Marc Berthier, Design Plan Studio
Alarm clock Pebble Objects
Aluminium jacketing, P.P.
h. 2cm (³/₄in) w. 6cm (2¹/₂in) l. 9cm (3¹/₂in)
Spirix-Lexon, France

Marc Berthier, Design Plan Studio
Data bank, Pebble Objects
Aluminium jacketing, P.P.
h. 2cm (³/₄in) w. 6cm (2¹/₂in) l. 9cm (3¹/₂in)
Spirix-Lexon, France

Marc Berthier, Design Plan Studio
Light, Pebble Objects
Aluminium jacketing, P.P.
h. 2cm (³/₄in) w. 6cm (2¹/₂in) l. 9cm (3¹/₂in)
Spirix-Lexon, France

Marc Berthier, Design Plan Studio
Pen set, Pebble Objects
Aluminium jacketing, P.P.
h. 2cm (³/₄in) w. 6cm (2¹/₂in) l. 9cm (3¹/₂in)
Spirix-Lexon, France

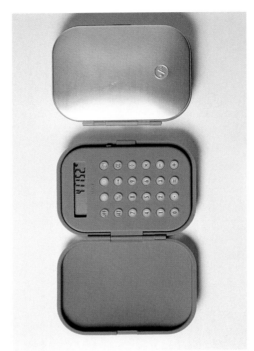

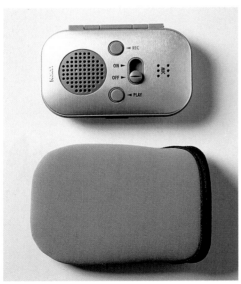

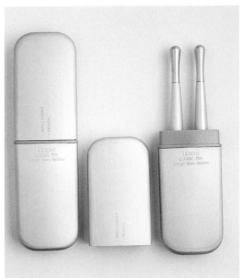

Geoff Hollington, Richard Arnott, David Townsend Elliot, Francis R. Skop Jnr.
Advanced photo system camera, Advantix T700
Various plastics
h. 6.5cm (2½in) w. 9.2cm (3⅝in) d. 3.6cm (1⅜in)
Eastman Kodak Company, USA

Hideki Kawai
Underwater camera, Ixus X-1
Polycarbonate, ABS plastic, stainless steel
h. 7.5cm (3in) w. 10.5cm (4⅛in) l. 4.7cm (1⅞in)
Canon Inc., Japan

Kaoru Sumita
Digital camera, DSC-F505 Cyber shot digital camera
Die cast magnesium
h. 6.2cm (2½in) w. 10.7cm (4¼in) l. 13.6cm (5⅜in)
Sony Corporation, Japan

Yasuhiko Miyoshi
Portable MD (Mini Disk) Player, MD-ST55(S)(A)(R)
Aluminium, ABS plastic
h. 7.8cm (3in) w. 7.1cm (2⁷⁄₈in)
Sharp Corporation, Japan

Sharp Corporation employs over 60,000 people in 33 countries with 66 factories based around the world. It is recognized internationally for products which emphasize technological standards while remaining both elegant and contemporary. The company was founded in 1912 by Mr Tokuji Hayakawa to produce metal objects following his patenting of a belt buckle, and the Sharp trademark derives from Hayakawa's invention of the 'Ever-Sharp Pencil'. Since these modest beginnings the company has developed into the No. 1 manufacturer of LCD (Liquid Crystal Display) and each year creates new markets by developing unique products from industrial electronics, to consumer electronic products and business equipment. Sharp, however, remains human-centric. It researches the customers' needs and produces designs which reflect these criteria. As our world becomes increasingly technological, and the equipment we use more involved, Sharp has recognized that, paradoxically, the user needs a simple interface between him/herself and that complexity. Its design philosophy is produce objects which have 'clarity and reliability' and which are 'friendly and enjoyable'.

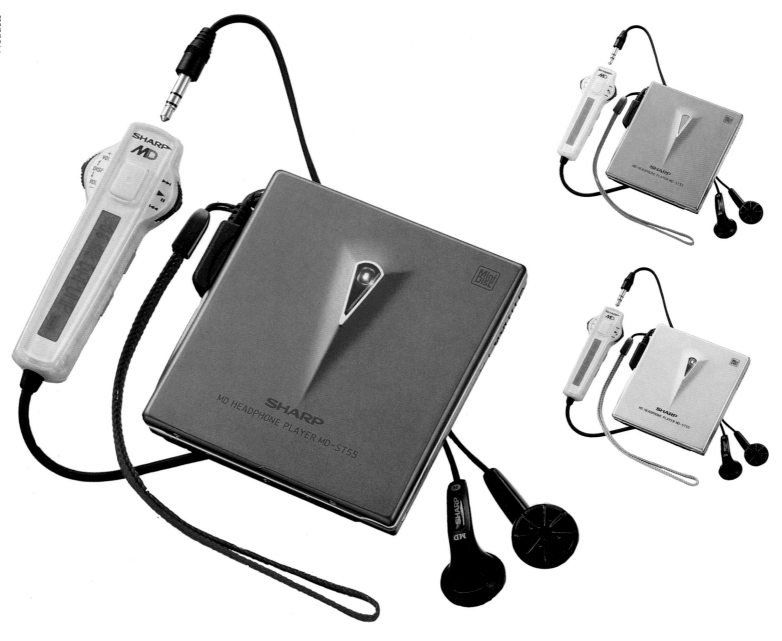

Kunihiro Okhi
Personal digital assistant, MI-C1-A
Plastic
h. 1.55cm (⅝in) w. 3.6cm (1⅜in) d. 8cm (3⅛in)
Sharp Corporation, Japan

Yumiko Takeshita
Facsimile, UX-E800
ABS plastic
h. 12.7cm (5in) w. 33.8cm (13⅜in) d. 26.5cm (10½in)
Sharp Corporation, Japan

Sachio Yamamoto, Katsunori Kume, Junichi Saitou
MPEG 4 Digital Video Recorder, VN-EZ5
Aluminium, plastics
h. 8.5cm (3⅜in) d. 4.82cm (1⅞in)
Sharp Corporation, Japan

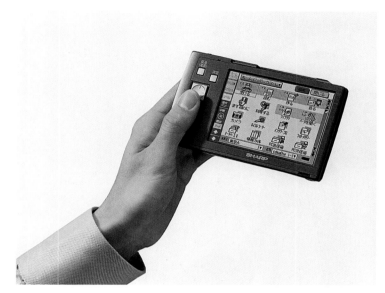

Philips
Earphones, HS 700 Outdoor Sport Earphones
Plastics
h. 10cm (4in) w. 5cm (2in) d. 1.5cm (½in)
Philips Electronics BV, The Netherlands

Products

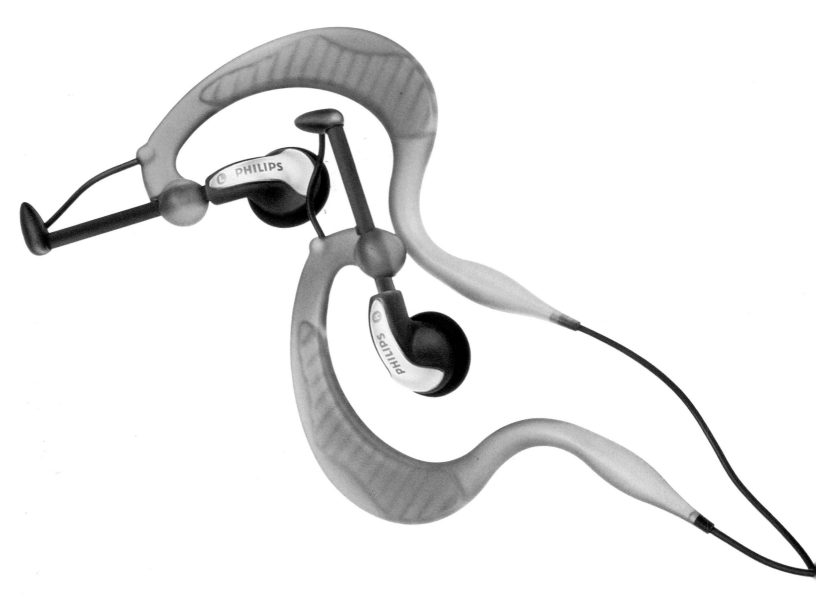

While internationally recognized as a producer of innovative electronics, Philips believes in 'humanware' and 'hardware'.

When the company first became involved in the mass production of consumer goods in the 1920s, its then Managing Director, Louis Kalff carried out market research to discover the differences in consumer's tastes worldwide so that these considerations could be added to Philips' already revolutionary designs. It wasn't until 1980 when Robert Blaich was made director of Philips Industrial Design Bureau that a policy was developed whereby design of any product was not only to be utilitarian and pragmatic but also address the more private and personal demands of the modern consumer.

Stefano Marzano, however (MD of Philips Design since 1991), is the creator of 'High Design' which takes this concept one step further. 'High Design' is a philosophy which is propagated throughout the whole company. Marzano believes that many human values have been lost from the world. Industrial design has become preoccupied with flashy gimmickry and less concerned with the needs of the consumer. He quotes Ezio Manzini of the Domus Academy – 'We have forgotten that objects are creatures produced by our spiritual sensibilities and by our practical abilities'. He considers that generally people are scared of technology and as MD of Philips Design he is responsible for allaying that fear. By using more traditional shapes and working with a research team of sociologists, anthropologists and psychologists to maintain a human focus, Philips has bridged the gap between technology and 'the man in the street'. His aim is to create a corporate identity which 'fathers meaningful objects that support people in their daily tasks, express the values they believe in, and stimulate their emotions and creativity'.

Philips
Temple thermometer, Sensor Touch Temple Thermometer HF 370
Plastic, metal
h. 17.5cm (6⁷⁄₈in) w. 4.3cm (1⁵⁄₈in) d. 3.3cm (¹⁄₄in)
Philips Electronics BV, The Netherlands

The 'Sensor Touch Temple Thermometer' is a revolutionary new infrared thermometer which measures temperature on the temporal artery. A gentle stroke of the sensor across the temple obtains an accurate reading. The design consideration was to combine innovative technology, a power source and a user feedback in a domestic product which was easy to use.

Philips
Residential phone, Kala™ Digital
Plastics
h. 16.3cm (6³⁄₈in) w. 5.3cm (2¹⁄₈in) d. 3.5cm (1³⁄₈in)
Philips Electronics BV, The Netherlands

The key to the design of the 'Kala' Digital residential phone was 'ease of use'; aimed at the humanizing of technology.

Philips
Skin care product, Cellesse SenseActive HP 5231
Neoprene
h. 16cm (6³⁄₈in) w. 11.6cm (4¹⁄₂in) d. 8.8cm (3¹⁄₂in)
Philips Electronics BV, The Netherlands

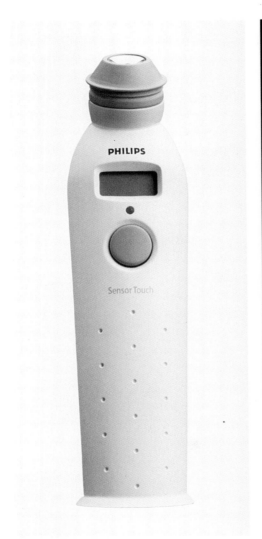

Frederick Lintz
Long range Walkie-talkie, Olympe
ABS plastic
h. 16.7cm (6½in) w. 3.7cm (1½in) l. 6.7cm (2⅝in)
Spirix Lexon, France

TKO Design
Mobile Phone, One Touch Easy DB™ Mobile Phone
ABS plastic, rubber, various polymers
l. 11.4cm (4½in) w. 5.2cm (2in) d. 2.8cm (1⅛in)
Alcatel Telecom, UK

Siemens
Mobile phone, S35I
ABS plastic
h. 2cm (¾in) l. 11.8cm (4⅝in) w. 4.5cm (1⅝in)
Siemens Electrogeräte GmbH, Germany

Siemens
WAP device, Unifier IC35
ABS plastic
h. 2cm (¾in) l. 10.8cm (4¼in) w. 8.7cm (3½in)
Siemens Electrogeräte GmbH, Germany

S35 is part of the new generation of Siemens mobile phones. It is palm-sized, watertight and durable. The IC35 unifier provides all the functions of the laptop and is connected to the internet via the handset. Music can be stored and listened to on headphones.

Claus-Christian Eckhardt
Mobile phone, Bosch 310
ABS-PC Silicon, chromium, stainless steel
h. 12.4cm (4⅞in) w. 4.4cm (1¾in) d. 2.4cm (1in)
Bosch Telecom GmbH, Germany

Claus-Christian Eckhardt
Mobile phone, Bosch 1886
ABS-PC Silicon, chromium, stainless steel
h. 11.2cm (4½in) w. 4.4cm (1¾in) d. 2.2cm (⅞in)
Bosch Telecom GmbH, Germany

Design 3 Produktdesign
Telephone, T-Easy C410 Cordless DECT Phone
Injection moulded ABS plastic
Handset: h. 14.8cm (5⅞in) w. 5.2cm (2in) d. 3.2cm (1¼in)
Base: h. 12cm (4¾in) w. 9.5cm (3¾in) d. 7.3cm (2⅞in)
Deutsche Telekom AG, Germany

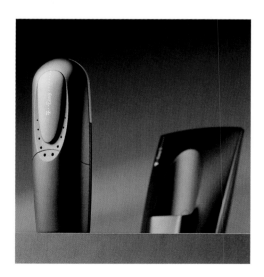

Hannes Wettstein
Watch, Futura
Titan-nitrogen
w. 4.5cm (1⁷/₈in) l. 2.5cm (1in) d. 0.9cm (³/₈in)
Ventura Design, Switzerland

Massimo Canalli
Monthly perpetual calendar, Endless
l. 42cm (16½in) w. 42cm (16½in)
Nava Design SpA, Italy

Hannes Wettstein
Watch, Ego Calculator
Titan-nitrogen
d. 0.9cm (³/₈in) di. 4.3cm (1⁵/₈in)
Ventura Design, Switzerland

Hannes Wettstein
Watch, Ego Chronometer
Titan-nitrogen
d. 0.9cm (³/₈in) di. 4.3cm (1⁵/₈in)
Ventura Design, Switzerland

Linde Design
Chronograph watch, Mexx City Chronograph
Stainless steel
di. 4.2cm (1⁵/₈in)
Mexx Time, Denmark

Apple Computer Inc.
Apple Cinema Display
Plastics
h. 47.9cm (18½in) w. 58.8cm (23in) d. 21cm (8½in)
Apple Computer Inc., USA

Apple Computer Inc.
Computer, Power Mac G4 Cube
Plastics
h. 24.8cm (9⁷/₈in) w. 19.5cm (7⁷/₈in) d. 19.5cm (7⁷/₈in)
Apple Computer Inc., USA

Apple Computer Inc.
Apple Power G4
Plastics
h. 43cm (17in) w. 22.6cm (9in) d. 46.7cm (18½in)
Apple Computer Inc., USA

Lunar Design
Computer Monitor, Sony Home Monitors
Injection moulded ABS plastic
h. 37cm (14⁵/₈in) w. 37.3cm (14³/₄in) d. 42.4cm (16³/₄in)
Sony Corporation, USA

Philips
Monitor Range, Philips 1999/2000 CRT Monitor Range
Plastics
h. 44.7cm (17⁵/₈in) w. 44cm (17³/₈in) d. 39.6cm (15⁵/₈in)
Philips Electronics BV, The Netherlands

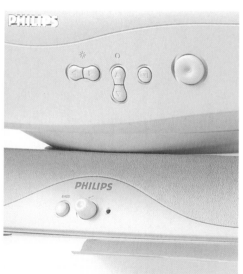

Philips
LCD Monitor, Philips Brilliance 151AX LCD Monitor
Acrylic
h. 41.8cm (16½in) w. 40.2cm (15⁷⁄₈in) d. 17.6cm (7in)
Philips Electronics BV, The Netherlands

Lunar Design
Home PC, HP Multimedia PC
ABS plastic
Tower: h. 37cm (14½in) w. 24.05cm (9½in) d. 35.5cm (14in)
Flat Panel: h. 44cm (17³⁄₈in) w. 39cm (15⁵⁄₈in) d. 21cm (8¼in)
Hewlett Packard Company, USA

Siemens
Mouse, ID Mouse
Plastics
Siemens Electrogeräte GmbH, Germany

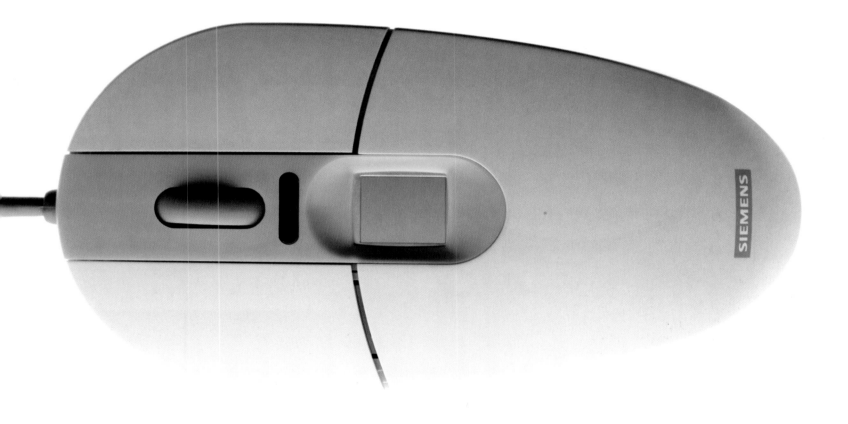

The ID Mouse recognizes the fingerprint of its owner. It is installed quickly under Windows but the computer's user has to first save his or her fingerprint as a reference. Sensors then create a representation of the print which is stored as a 'visual' password.

'Siemens – we make knowledge work for you' is the motto Siemens used to present itself as a knowledge-based company of the 21st Century at Expo 2000 in Hanover. Since the company was founded in 1847 it has built up a reputation for innovation in the fields of electrical engineering, information and communication systems, health care and, more recently, environmentally-friendly power generation and electronics.

Siemens AG (set up in 1966) has developed 80 per cent of its electronic products over the past five years confirming its stated goal of constantly re-appraising and refining its product range – a programme supported by a worldwide R&D team of 50,000. The division recently merged with Publicis MCD Messeagentur to form Siemens Design and Messe GmbH, headed by Herman Schultes, with the aim that the two departments should work as a team.

Its design philosophy is to produce goods which combine functional clarity with a rational aesthetic and its decision to involve renowned designers to work as consultants to provide new ideas and stimuli means that recent designs have become more multifaceted and wide-ranging. One of its main objectives is not only to produce ground-breaking and profitable goods but in so doing not to lose sight of the broader picture and the overall benefits to society. It spends billions of DM each year developing environmentally friendly production techniques that cut down on the use of energy and materials and reduce emissions.

Mitsuhiro Nakamura
15th Anniversary CD walkman, D-EJ01
Magnesium alloy
h. 2.15cm (⁷/₈in) l. 13.85cm (5½in) w. 13.62cm (5³/₈in)
Sony Corporation, Japan

Noriaki Itai
DVD Player with 7in wide LCD monitor, DV-L70S
Plastic
h. 2.54cm (1in) w. 18.8cm (7in) d. 14.1cm (5½in)
Sharp Corporation, Japan

Shinichi Obata
VAIO Music Clip, MC-P10
Plastic mould
h. 2cm (⁷/₈in) w. 12cm (4⁷/₈in) d. 2cm (⁷/₈in)
Sony Corporation, Japan

The MC-P10 Music Clip stores and plays digital music downloaded from the net or transferred from an audio CD. A high-speed music file transfers an hour of music in approximately three minutes.

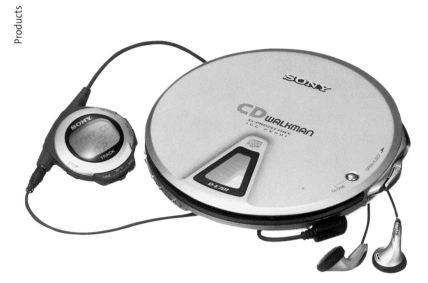

Yoshinori Inukai
Colour flatbed scanner, CanoScan FB636U
Aluminium, ABS plastic, glass
h. 3.9cm (1½in) w. 25.6cm (10in) l. 37.25cm (14⅝in)
Canon Inc. Design Centre, Japan

Yoshinori Inukai
Colour flatbed scanner, CanoScan FB330P/FB630P
ABS plastic, glass
h. 3.9cm (1½in) w. 25.6cm (10in) l. 37.25cm (14⅝in)
Canon Inc. Design Centre, Japan

Airi Itakura
Colour flatbed scanner, CanoScan FB1210U
ABS plastic, aluminium
h. 9.25cm (3⅝in) w. 28.6cm (11¼in) l. 46.1cm (18⅛in)
Canon Inc. Design Centre, Japan

Oba Haru
VAIO PC, PCV-L720
ABS plastic
LCD display: h. 33cm (13⅞in) w. 43cm (16⅞in) d. 17.5cm (6⅞in)
VAIO Smart Keyboard: h. 3.5cm (1⅞in) w. 38cm (15⅞in)
d. 17.5cm (6⅞in)
Sony Corporation, Japan

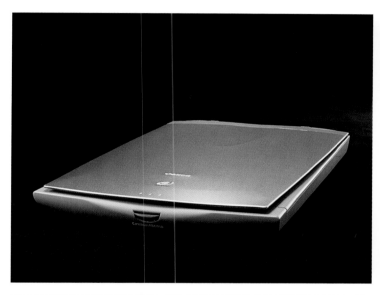

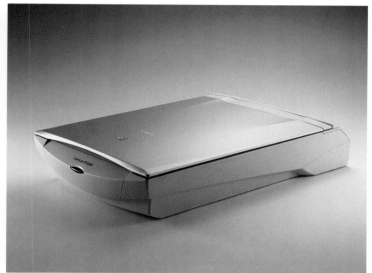

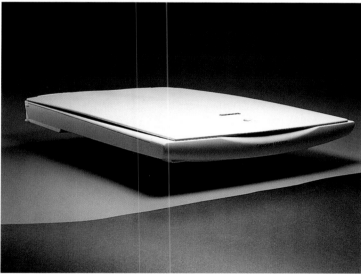

The angle of PCV-L720's multimedia LCD display has a greater range of moveability compared to other LCD displays, thanks to the dual-hinged pedestal. Oba Haru wanted to design a computer that could be used in a relaxed manner, like reading a book.

Toshiyuki Kita
28in TFT LCD TV, LC-28HD1
Aluminium
h. 44.8cm (17⁵/₈in) w. 69.08cm (27¹/₈in) d. 6cm (2³/₈in)
Sharp Corporation, Japan

Philips
Television, New Vision 1
h. 37cm (14¹/₂in) w. 41cm (16¹/₈in) d. 38cm (15in)
Philips Electronics BV, The Netherlands

Philips
Television, FL-7 Design Line Silver Gloss 32in widescreen TV
Plastics
h. 104cm (41in) w. 92cm (36¹/₄in) d. 60cm (23⁵/₈in)
Philips Electronics BV, The Netherlands

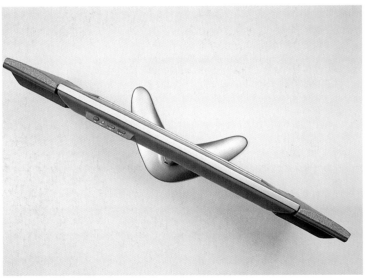

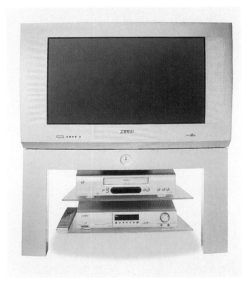

The LC-28HD1 is the world's first 28in wide screen LCD television compatible with digital hi-vision broadcasts.

The New Vision 1 television fulfils a consumer need. The 14in TV has been developed through market research: the needs and aspirations of a chosen segment of society were identified and a bedroom television was designed which also contains analogue clock and ambient lighting.

Yamaha Product Design Laboratory
Grand Piano, Yamaha Disklavier pro 2000
Spruce, cherry, maple, aluminium, iron
h. 102cm (40⅛in) w. 155cm (61in) d. 227cm (89⅜in)
Yamaha Corporation, Japan
Limited batch production

NAC Sound
Speaker, Atun
Ceramic
h. 74cm (29in) w. 20cm (7⅞in)
NAC Sound srl, Italy

NAC Sound
Speaker, Seth
Ceramic
h. 32cm (12½in) w. 14cm (5½in)
NAC Sound srl, Italy

NAC Sound
Speaker, Isis
Ceramic
h. 74cm (29in) w. 20cm (7⅞in)
NAC Sound srl, Italy

NAC Sound
Speaker, Zemi
Ceramic
di. 24cm (9½in)
NAC Sound srl, Italy

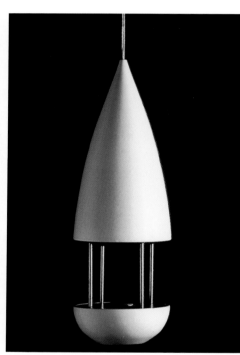
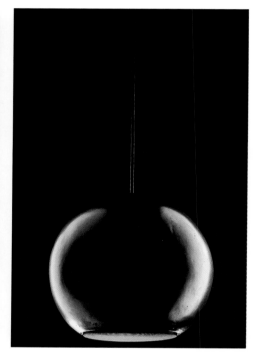
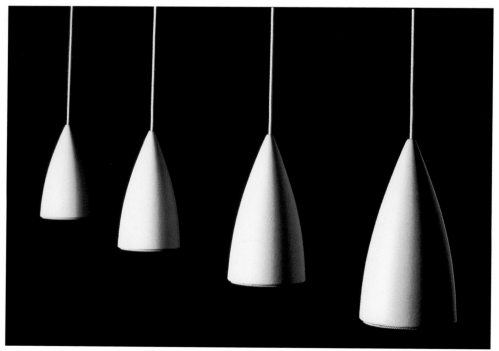

NAC Sound combines innovative design with the latest in speaker development. Its products emit spherical rather than conical sound and its patented acoustic concept, 'Omnidirectional Technology' means that speakers can be placed anywhere in the room. The original designs were mainly in ceramic but today the group is experimenting with materials as diverse as aluminium, carbon fibre and wood. A NAC product is instantly recognizable by its shape. Long and tapered or spherical, the choice is not merely an aesthetic one; these forms are less likely to suffer any mechanical vibrations and unwanted resonance.

AVC Design Centre
Portable DVD player with LCD, DVD-LA75
Plastic
h. 24.8cm (9¾in) w. 18.5cm (7¼in) l. 14cm (5½in)
Matsushita Electrical Industrial Co. Ltd, Japan

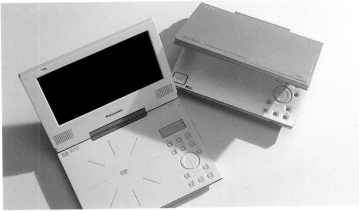

Nobuhiro Fujii
Upright type cyclone vacuum cleaner
Plastic, rubber
h. 102cm (40⅛in) w. 28.4cm (11⅛in) d. 16.5cm (6½in)
Sharp Corporation, Japan

Luigi Molinis
Fan coil, Frend
ABS plastic
h. 45cm (17¾in) w. 76.5cm (30⅛in) d. 17.5cm (6⅞in)
Rhoss SpA, Italy

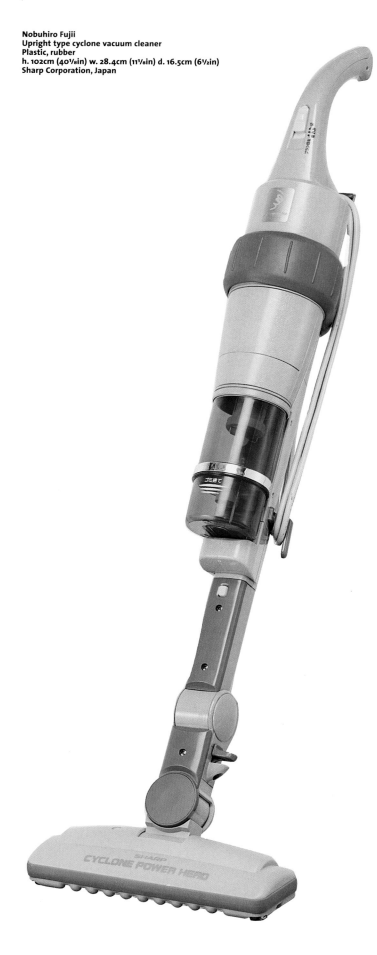

Tomoki Taira
Microwave oven, Half Pint R-120D
ABS plastic, steel
h. 36cm (14⅛in) w. 36.5cm (14⅜in) d. 36.2cm (14¼in)
Sharp Corporation, Japan

The Half Pint R-120D microwave is a 'life-style' appliance. It comes in a range of colours and its compact size makes it ideal for use in the home/office and other space-restricted environments. The internal capacity, however, has not been compromised so it is equally adaptable to the family kitchen.

Carsten Joergensen
Electric coffee maker, Santos
Polycarbonate
h. 32cm (12⅝in) di. 18cm (7in)
Bodum AG, Switzerland

Massimo Iosa Ghini
Coffee machine, 'T'
h. 41cm (16⅛in) w. 28cm (9⅞in) d. 24cm (9½in)
Tuttoespresso, Italy

Kazuhiko Tomita
Teapot, Ciacapo
Cast iron
h. 13cm (5⅛in) w. 18cm (7in) di. 14cm (5½in)
Covo srl, Italy

Philips
Coffee maker, Café Duo HD 7140/42
Plastic, metal
h. 25cm (9⅞in) w. 17cm (6¾in) d. 17cm (6¾in)
Philips Electronics BV, The Netherlands

Stefano Giovannoni
Jug, Alibaba
PMMA glass
h. 25cm (9⅞in) di. 17.5cm (6⅞in)
Alessi SpA, Italy

Ann Morsing, Beban Nord
Magazine and paper basket, Pelle box
Fibreboard
h. 30cm (11⅞in) w. 43cm (17in) d. 31cm (12¼in)
Box Design AB, Sweden

Henrik Holbaek, Claus Jensen
Ice cube cooler, Eva Solo
Stainless steel, plastics
h. 7cm (2³⁄₄in) di. 17cm (6⁵⁄₈in)
Eva Denmark A/S, Denmark

Henrik Holbaek, Claus Jensen
Vacuum flask, Eva Solo
Stainless steel, plastics
h. 27cm (10⁵⁄₈in) di. 13cm (5¹⁄₈in)
Eva Denmark A/S, Denmark

Henrik Holbaek, Claus Jensen
Citrus squeezer, Eva Solo
Glass, stainless steel, plastics
h. 18cm (7in) di. 11cm (4³⁄₈in)
Eva Denmark A/S, Denmark

Henrik Holbaek, Claus Jensen
Knife Stand, Eva Solo
Aluminium and plastics
h. 26cm (10¹⁄₄in) w. 7.5cm (3in)
Eva Denmark A/S, Denmark

Sam Hecht, Naoto Fukasawa, IDEO
Matsushita cooking tools collection
ABS, porcelain, tin plate, stainless steel, tempered glass

Water Kettle
h. 16cm (6¼in) w. 20cm (7⅞in) d. 24cm (9½in)

Egg Steamer
h. 16cm (6⅜in) w. 5.7cm (2¼in) d. 5.7cm (2¼in)

Juicer
h. 10cm (4in) w. 11cm (4⅜in) d. 11cm (4⅜in)

Rice Maker
h. 20.5cm (8in) w. 20cm (7⅞in) d. 24cm (9½in)

Toaster
h. 15cm (6in) w. 18cm (7⅛in) d. 16.5cm (6½in)

Coffee Maker
h. 25cm (9⅞in) w. 9.5cm (3¾in) d. 9.5cm (3¾in)

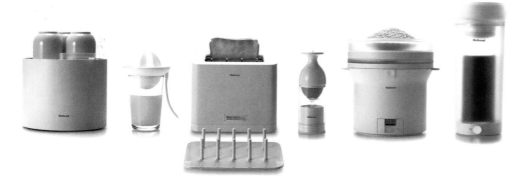

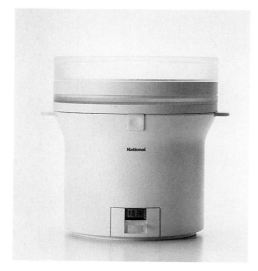

Matsushita developed its cooking tools collection as a reaction to the changing lifestyle of the average Japanese family. Breakfast is now a hybrid of western and eastern traditions – a rich selection of foods often eaten 'on the go'. The products to cook them, however, have not been available. The company's brief for the series was that the utensils should be small and compact, operate in a variety of ways, be useable in different settings (kitchen or table) and have a reduced amount of parts to help with food preparation, serving and cleaning.

The Rice Maker is the first rice cooker that can also grind; the two parts can be disassembled and used independently.

The size of the Juicer mechanism has been reduced to a minimum and resembles the shape of an orange. The juice is delivered into the same glass from which it is to be drunk.

By using steam, an egg can be cooked in the Egg Steamer to the desired softness, depending on how much water is used. The lid contains the steam and also acts as the egg cup.

Paolo Ulian
Kitchen knife, Pane e Salame
Steel, aluminium, wood
h. 4cm (1½in) w. 1.5cm (⅝in) l. 33cm (13in)
Zani & Zani, Italy

Paolo Ulian, Guiseppe Ulian
Pizza knife, Rotella Tagliapizza
Steel
h. 8cm (3⅛in) w. 2cm (⅞in) l. 26cm (10¼in)
Zani & Zani, Italy

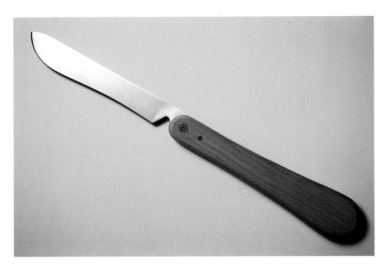

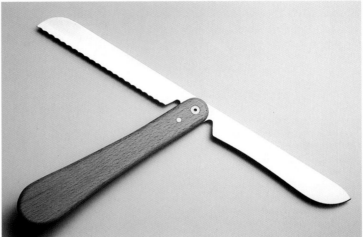

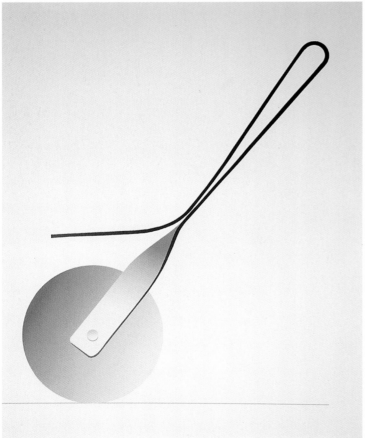

Asher Stern
Drinks chess set, Red vs. White
Wood, aluminium, glass
l. 44cm (17³/₈in) w. 44cm (17³/₈in)
Limited batch production

Jürgen Schmidt
Cordless Screwdriver
w. 6cm (2³/₈in) l. 20cm (7⁷/₈in) d. 6cm (2³/₈in)
Metabowerke, Germany
Prototype

Tobias Koeppe
Paper scissors, Cassini
Zinc, steel, ABS plastic, PMMA
h. 0.85cm (³/₈in) w. 6.2cm (2¹/₂in) l. 21cm (8¹/₄in)
Lerche, Germany

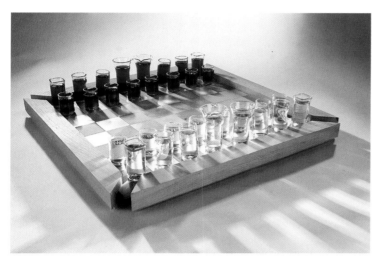

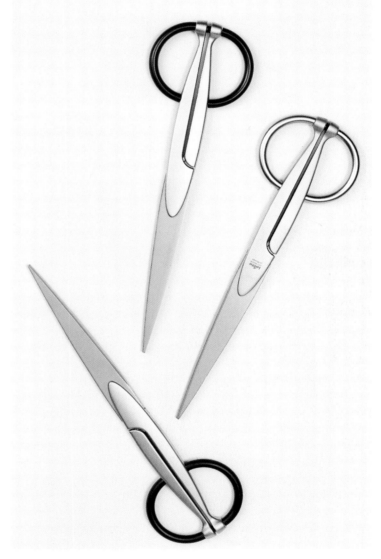

Philips
Shaver, Philishave Quadra Action 6000 Series
Steel
h. 14.5cm (5¾in) w. 5cm (2in) d. 7cm (2¾in)
Philips Electronics BV, The Netherlands

Shinichi Sumikawa
Toothbrush, Swan
Plastic
h. 20cm (7⅞in) w. 1cm (⅜in) d. 15cm (6in)
Iridium, Korea

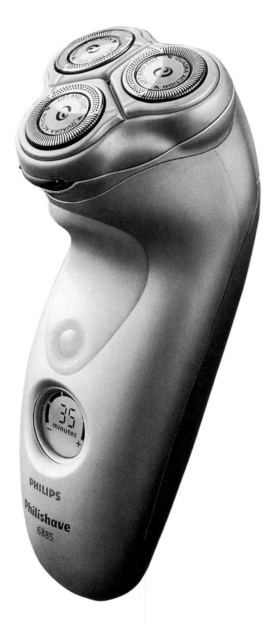
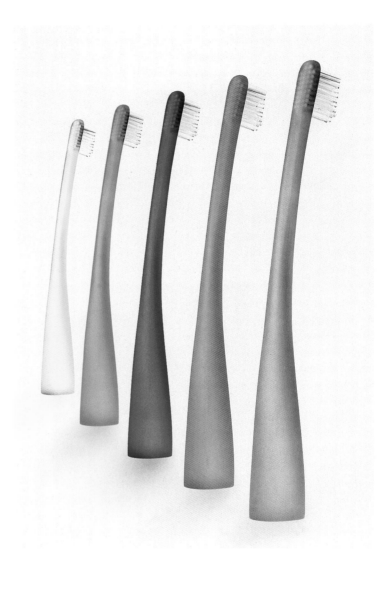

Ippei Matsumoto
Sponge Scale
Sponge, LED, plastic
h. 5cm (2in) w. 29cm (11³/₈in) d. 25cm (9⁷/₈in)
IDEO, Japan

Makoto Hashikura, IDEO Japan
Keyboard mousepad
Neoprene rubber, sponge, ABS plastic
Matsushita Electric Corporation, Japan

Keyboard mousepad
h. 1.2cm (½in) w. 38cm (15in) d. 16cm (6³/₈in)

Mouse
h. 2.5cm (1in) w. 8cm (3in) d. 2.5cm (2in)

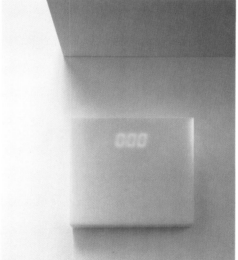
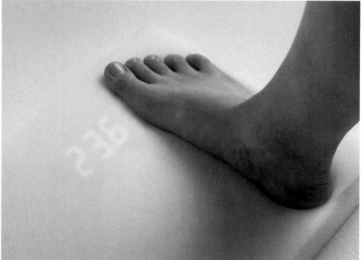

In the Autumn of 1999 IDEO, Japan, and the Diamond Design Management Network staged a series of workshops for designers from several companies in various industries to explore the way people react emotionally to and communicate intuitively with the products that surround them. The resulting designs were exhibited in a show entitled '"Without Thought" – design for the subconscious'. Each object plays on the memories and feelings that we all share and is created to appeal to our senses.

The sponge scales are soft and respond to one's weight. For the more sensitive minded, the display remains fuzzy.

The mouse pad itself is a key board so that one can both type and point using the same work surface.

Luigi Trenti
Writing Instruments, Collezione Egosphere
Resin, steel, gold/platinum plated
Various sizes
Francesco Pineider SpA, Italy

dai design
Rollerpen, Ihag
Aluminium, stainless steel
l. 14cm (5½in) di. 1cm (³/₈in)
Privatbank IHAG Zurich AG, Switzerland

Shibuyo Ito, Setsu Ito
Pen stand, How
Plastic
h. 10cm (4in) w. 8.5cm (3³/₈in) l. 8.5cm (3³/₈in)
Nava Design SpA, Italy

Shibuyo Ito and Setsu Ito
Pen holder, How
Plastic
h. 7cm (2³/₄in) w. 10cm (4in) l. 10cm (4in)
Nava Design SpA, Italy

Shibuyo Ito, Setsu Ito
Sottomano, How
Plastic
w. 70cm (27¹/₂in) l. 50cm (19⁵/₈in)
Nava Design SpA, Italy

Shibuyo Ito, Setsu Ito
Sottomano, How
Plastic
w. 70cm (27¹/₂in) l. 50cm (19⁵/₈in)
Nava Design SpA, Italy

Shibuyo Ito, Setsu Ito
Scotchtape holder, How
Plastic
h. 7cm (2³/₄in) w. 10cm (4in) l. 10cm (4in)
Nava Design SpA, Italy

David Farrage, Goeran Jerstroem
Digital thermometer, Vicks Comfort Flex Digital Thermometer
ABS plastic, Kryton rubber, acrylic
w. 0.3cm (⅛in) l. 13cm (5⅛in)
OXO International, USA

Smart Design
Text messager, Thumbscript Concept Prototype Communicator
Silicone, urethane, metallic finish
h. 7.6cm (3in) w. 5.3cm (2⅛in) l. 1.9cm (¾in)
OXO International, USA
Prototype

David Farrage, Vanessa Sica, Kevin Lozeau
Humidifier, Kaz Health Mist Humidifier
Polypropylene
h. 15.24cm (6in) w. 33cm (13in)
OXO International, USA

David Farrage
Scrub brushes, OXO Good Grips Scrub Line
Polypropylene, santoprene, nylon
Various sizes
OXO International, USA

David Farrage, Dean Chapman, Davin Stowell
Palmtop communications and medications dispensing device, Medimonitor
Plastic, LCD screen
h. 3.6cm (1⅜in) w. 22.8cm (9in) l. 12.2cm (4⅞in)
OXO International, USA

Kazuyo Komoda
Children's toy, La Tavola Imbandita ('Lair')
Cotton, foamed polyurethane, aluminium, wood
h. 50cm (19⅝in) w. 135cm (53⅛in) d. 80cm (31½in)
Edition Galleria Luisa delle Piane, Italy

Lamberto Angelini
Suitcase, Frog
PPL (polypropylene) injected
Large: h. 62cm (24⅜in) w. 27cm (10⅝in) l. 80cm (31½in)
Small: h. 54cm (21¼in) w. 24cm (9½in) l. 69cm (27⅛in)
Roncato SpA, Italy

Design Team Mandarina Duck
Trolley, Frog Jr.
Polypropylene, fabrics
h. 55cm (21⅝in) w. 40cm (15¾in) d. 20cm (7⅞in)
Mandarina Duck Plastimoda, Italy

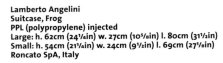

Kazuyo Komoda's 'Lair' formed part of the show mounted by Luisa delle Piane in Milan this year. The collection of children's furniture by designers, including Konstantin Grcic and Matali Crasset, sought to invent a world through a child's eye and not just a reduction of adult fantasy.

TKO Design
Washing machine, Titan
Steel, polymers
h. 85cm (33½in) w. 60cm (23⅝in) d. 60cm (23⅝in)
Monotub Industries, UK
Prototype

Jam Design
Concept kitchen, Corian
Various sizes
Jam Design and Communications Ltd,
UK

Christophe Pillet
Kitchen system, Aerosystem
Stainless steel, glass, walnut, aluminium
Various sizes
Ciatta a Tavola, Italy

The young British design team Jam collaborated with major product manufacturers Corian, Whirlpool, Softroom and Linbeck Rausch Lighting to conceive a concept kitchen which was exhibited at 100% Design in London. This installation demonstrated that tomorrow's kitchen will be smart, highly automated, multifunctional and flexible; mundane tasks will be removed to leave time for more expressive activities. The design consisted of a highly technical central piece used for refrigeration, storage, food preparation, eating and cooking while Whirlpool products showed the way in which temperature regulations, lighting and electronic control systems could develop. However, the Corian finish to the work surfaces, Linbeck lighting and use of soft curves create a sensual environment suggesting a stronger emotional role within new living spaces.

Merloni Elettrodomestici
Home computer, Leon@rdo
Novakval ABS plastic 5V
h. 20cm (8in) l. 25cm (10in) w. 35cm (13³⁄₄in)
Ariston, Italy

The future for domestic technology lies in WRAP (Web Ready Appliances Protocol). With a networked home it is possible to communicate with electrical appliances and the outside world through any telecommunications network, and with Leon@rdo it is now also possible to communicate through the internet. The Ariston Digital Service Centre monitors the energy consumption in a household and warns the occupier of any problem; similarly household appliances can self-diagnose any malfunctions and notify the centre. As Leon@rdo is a home computer one can also e-mail and order groceries and download new recipes directly to the oven (e-cooking).

Sacha Winkel
Wash basin, Soap
Ekotek
h. 33cm (13in) l. 80cm (31½in) d. 45cm (17¾in)
Boffi, Italy

Claudio Silvestrin
Kitchen, Xila
Laminates, setasil, HPG, polyester
Various sizes
Boffi, Italy

Val Cucine
Kitchen, Ricicla Laminato
Aluminium, laminates, wood
Various sizes
Val Cucine SpA, Italy

Claudio Silvestrin
Wash basin, Adda
Ceramic
h. 17.5cm (7in) di. 50cm (19½in)
Boffi, Italy

The Val Cucine company policy is to respect the environment. Its kitchens are made from natural materials and all its production technologies use substances that have been tested to guarantee maximum recycling capacity. Taking its philosophy one stage further, it is closely linked with 'Bioforest', an association founded to promote regeneration of natural environments. It has recently financed two reforestation programmes in Brazil and Ecuador and its goal is to restore more forests in order to absorb the carbon dioxide produced by the manufacturing of furniture.

Norbert Wangen
Kitchen, Cube 2000
Stainless steel, wood
h. 93cm (36⁵/₈in) w. 70cm (27¹/₂in) l. 234cm (92¹/₈in)
Norbert Wangen, Germany

Norbert Wangen's compact cube contains all that is necessary in a kitchen: fridge, dishwasher, oven/microwave, sink and storage space. The moveable work surface can be pushed along to reveal the sink and cooker and acts as a table. 'A complete kitchen that can appear and disappear before your eyes,' says Wangen.

Bulthaup
Kitchen, System 25
Stainless steel, linoleum
Various sizes
Bulthaup, Germany

Bulthaup
Kitchen, System 20
Stainless steel, aluminium alloy
Various sizes
Bulthaup, Germany

Giampaolo Bendini
Bathtub, Spoon
Exmar (marmor & resins)
h. 48cm (18⅞in) w. 101cm (39¾in)
Agape srl, Italy

Giampaolo Bendini
Tap, Fez 40
Chromed brass
Agape srl, Italy

Giampaolo Bendini
Tap, Fez 260
Chromed brass
Agape srl, Italy

Maximillian Burton
Toilet brush, OXO good grips toilet brush
Polypropylene, santoprene, nylon bristles, ABS plastic
h. 45.7cm (18in) w. 14cm (5½in)
OXO International, USA

Stefano Giovannoni
Plunger, Johnny the Diver
Thermoplastic resin
h. 26cm (10¼in) di. 13.3cm (5¼in)
Alessi SpA, Italy

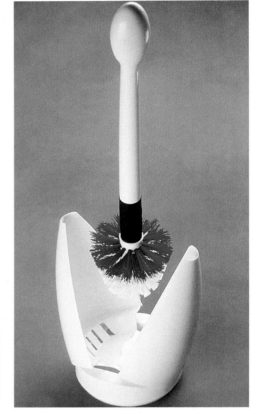
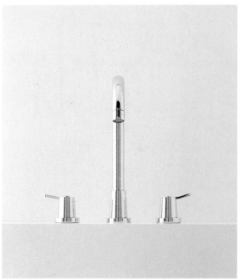

Roderick Vos
Ceramic kitchenware, Bricks for the House
Ceramic with glaze
h. 11.4cm (4½in) w. 11.4cm (4½in) l. 11.4cm (4½in)
Driade SpA, Italy
Limited batch Production

Martin Szekely
Flower pot, Brique à Fleurs à Vallauris
Clay
h. 52.5cm (20⅝in) di. 46.5cm (18⅜in)
Galerie Kreo, France

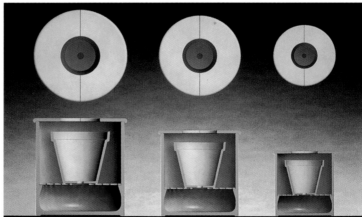

Emma Quickenden
Radiator, Hot Water Bottle
Rubber
h. 76cm (30in) l. 51cm (20in)
Prototype

Emma Quickenden
Radiator, Fire Place
Pressed Steel
h. 76cm (30in) w. 51cm (20in)
Prototype

Dick Van Hoff
Concrete stove
Concrete, metal
h. 100cm (39⅜in) w. 52cm (20½in) l. 80cm (31½in)
Van Hoff Omptwerper, The Netherlands
Prototype

Emma Quickenden has 'warmed' our radiators. She has reinvented their typology – the ugly white metal boxes have been replaced by designs redolent with nostalgia and cosiness and, as such, she hopes that they will become focal points within a room. The 'Hot Water Bottle' is made from rubber. Water flows through metal pipes within the flexible form which allows one to hold, squeeze, lean on and even smell it.

BMW
C1

What is it? Is it a bike with a roof or a car without sides? Whatever, this hybrid has been adapted for the urban commuter who would like the freedom that a two-wheeled vehicle brings to weave through the traffic, cutting journey times considerably, with the safety that a good strong roof over one's head promises. Safety was one of the major considerations for BMW when it developed its C1 motorbike (we all know that's what it really is). The roof has a roll bar and is crash tested; the engine won't start unless the seat belt has been engaged; there is a projection over the front wheel which is designed to mitigate any head-on collision; there are side-impact bars around the shoulders; and any shock is designed to be taken by the frame rather than the body. It is a captivating piece of design, having the similar head-turning quality of the Smart car when it first appeared on our city streets. But what are its drawbacks? Aidan Walker, managing editor of *Blueprint* and a committed biker, has written that it has one or two disconcerting features. Because of its towering frame, when the bike leans as it corners the rider feels thrown off balance. Secondly, the front of the bike is wide, limiting forward vision – it is impossible to see what the front wheel is doing. Ultimately, however, these are problems which can be overcome with use. BMW's new design has successfully developed a new typology for the road.

Marc Newson
Bicycle, Biomega
Super formed aluminium
h. 67cm (26³/₈in) w. 17cm (6⁷/₈in) l. 83cm (32⁵/₈in)
Magis srl, Italy

Stile Bertone SpA
Mountain bike, MTB fully suspended
Carbon fibre magnesium
h. 100cm (39in) w. 56cm (22in) l. 175cm (68in)
Stile Bertone SpA, Italy

Honda
Concept car, Neukom

Honda
Concept car, Fuya-jo ('Sleepless City')

We are witnessing a period of huge transformation in the external shapes and interior environments of cars. In this design evolution we see the most advanced visions of a new world emerging. The innovations in shape, style and mobility produce new concepts for living in both the cities and countryside, and the passenger space becomes either a mere compartment for moving around in or an extra living room. (MDL)

The Neukom is described by Honda as a communication capsule. The wrap-around glass offers a full panorama while the swivelling seats create a meeting space on wheels.

The Fuya-jo or 'Sleepless City' was conceived for the young and fun-loving driver. Occupants riding in a semi-standing position are meant to be offered the same thrilling sensation experienced on a skateboard or on roller blades. With a steering wheel like a turntable, an instrument panel resembling a DJ's mixer and powerful amplifiers in the doors, it is a mini-nightclub on wheels.

Christopher C. Deam, Wilsonart International, Inside Design
Caravan
Aluminium (exterior), laminates (interior)

Fiat
Concept car, Eco-Basic

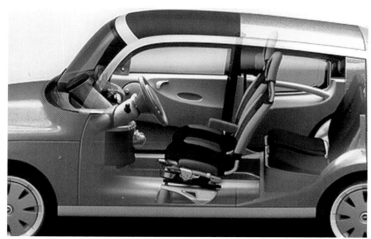

This caravan is a collaboration between Wilsonart International (manufacturer of decorative surfacing products); Inside Design (a company dedicated to the promotion of contemporary design through product development, special events and educational media) and Christopher Deam (one of the new wave of American designers who is presently receiving international recognition). The aluminium trailer has long been an indelible symbol of the American good-life. This 1948 example, which was created as the centrepiece of the 2000 International Contemporary Furniture Fair in New York, has been given a new lease of life. Jim Huff of Inside Design believes the interiors of these icons have never been as pleasing as the exteriors and he wanted to bring them into line. Deam used laminate to form an inside shell as it is lightweight, durable, affordable and warm in feel, contrasting well with the coolness of the aluminium. Long associated with prefabricated housing, it was a natural choice – 'nostalgically familiar yet modern, even futuristic in attitude' (Grace Jeffers, design historian and consultant to Wilsonart). The result is a design classic reinvented to appeal to the younger aesthetic.

With an engine designed to be economical and environmentally friendly, the Eco-Basic consumes only 3 litres of diesel per 100 kilometres. Its sleek aerodynamic shape and a 'start and go' strategy (this causes the engine to be immobilized after 4 seconds of standstill and restarted by a touch of the accelerator) make it the ideal compact car for the city of tomorrow.

Biographies

Lodovico Acerbis was born in Italy in 1939 and studied Economics and Business Studies at Milan University. In 1963 he joined the family firm of which he is now president. Acerbis International SpA was founded in 1870 as a carpenter's shop but it quickly expanded to industrial proportions and became one of the first companies to use architects for furniture design. Acerbis has earned a reputation for blending modern design with avant-garde technology, working with the finest materials and best designers. A number of Acerbis products have been awarded the Milan Compasso d'Oro Award, and are on permanent display in the Victoria and Albert Museum, London; Die neue Sammlung Staatliches Museum fur Angewandte Kunst, Munich; the Museum of Contemporary Art, Chicago; and in the City Hall, Shanghai. 49

Werner Aisslinger was born in 1964 in Berlin. He founded his own design company in 1993 and since then has carried out various furniture projects for Italian companies such as Magis, Cappellini and Porro. His 'endless shelf' won the Bundespreis Produkt design and in 1997 his 'Juli' chair for Cappellini was acquired as part of the permanent design collection at MoMA, New York. Aisslinger has also carried out corporate architecture concepts and projects for Lufthansa and Mercedes Benz. 43, 48, 64

Gunilla Allard was born in 1957. She worked as a stage designer and property manager for the film industry during the early 1980s before attending the University College of Arts, Crafts and Design in Stockholm to study interior architecture. Today she continues to work as a stage designer and is also well-known in Sweden for furniture design. She has been awarded the Excellent Swedish Form prize in 1991, 1992, 1996 and 1999. 33

Harry Allen received a Masters in Industrial Design from the Pratt Institute, New York. He worked for Prescriptive Cosmetics before opening his own studio. He immediately received success with his 'Living Systems' furniture line which he showed at the ICFF in New York and which led to commissions including the Sony Plaza and the North Face Store in Chicago. He has also earned international recognition for his interior design projects most notably the Murray Moss shop in New York – which he is currently expanding – and the 'Dragonfly Selects' jewellery store in Taiwan. He has recently completed new offices for *Metropolis* magazine and for the Guggenheim Museum, New York. Product designs include a medicine chest for Magis, various articles for Wireworks and lighting designs for George Kovacs and IKEA. His use of innovative material can be seen in his series of ceramic foam lamps which are on display in MoMA, New York and which he showed in the Mutant Material in Contemporary Design exhibition in 1994. 61

Emilio Ambasz was born in 1943 in Argentina but studied at Princeton University after which he took up a post as curator of design at MoMA, New York. From 1981 to 1985 he was President of the Architectural League and taught at Princeton's School of Architecture. He has also been visiting professor at Ulm Design School in Germany. In 1992 he was awarded first prize at the Universal Exhibition, Seville and his works are represented in museums worldwide, including MoMA and the Metropolitan Museum, New York. In 1989 there was a major retrospective of his architectural work at MoMA. 29, 33

Yoko Ando was born in Shibuya, Tokyo. He trained as a graphic artist at the Musashino Art University and is currently working for NUNO Corporation. He has exhibited work within Japan and at the 'L'Asie en Rose' Fashion show in Manila, the Philippines. 177

Lamberto Angelini founded Angelini Design in Bologna in 1980. A graduate in mechanical engineering, he worked for two years at the Volkswagen Style Centre, Wolfsburg, Germany, and from there moved into product design. The first Angelini products were all vehicles but the company now works across many areas for a diverse range of clients such as Acquaviva, Ducati, Motobechane, BMW, Castelli, Fogacci, Marcato, Mandarina Duck and Goldoni. 210

Ron Arad was born in Tel Aviv in 1951. He studied at the Jerusalem Academy of Art, and from 1974 to 1979 at the Architectural Association, London. In 1981 he founded One Off with Caroline Thorman and in 1983 designed their first showroom in Neal Street, Covent Garden, London. Early pieces include the Rover Chair, the vacuum-packed Transformer Chair and the remote-controlled Aerial Light. Later work explored the use of tempered steel, in the Well-Tempered Chair and the Bookworm. In 1988 Arad won the Tel Aviv Opera Foyer Interior Competition with C. Norton and S. McAdam, and formed Ron Arad Associates with Caroline Thorman and Alison Brooks in order to realize the project. As well as the design and construction of the new One Off design studio, furniture gallery and workshop in 1991, projects have included furniture design for Poltronova, Vitra, Moroso and Driade; the design of various interior installations; the interiors of the restaurants Belgo and Belgo Centraal in London, domestic architectural projects and the winning competition entry for the Adidas Stadium, Paris (unbuilt). In 1994 Arad established the Ron Arad Studio in Como, Italy, to expand on the production of limited edition handmade pieces. Arad was a guest professor at the Hochschule in Vienna from 1994 to 1997 and is currently Professor of Furniture and Product Design at the Royal College of Art in London. In 2000 his work was the subject of a major retrospective at the Victoria and Albert Museum, London. 31, 35, 141

Architettura Laboratorio is an industrial design group founded by Andrea Cacai, Maurizio Giordano and Roberto Grossi. These three architects have worked together and exhibited at Milan since 1992. They have collaborated with Colombo Design, Gruppo IPE and Gruppo Scaroni. 88

Neil Austin was born in 1958 and studied furniture design at Buckinghamshire College, England, where he is now a course leader in Furniture and Related Product Design. After graduation he worked as a professional draughtsman, and was then Design Director for a London furniture company. He regularly exhibits in conjunction with Mo-billy and has shown his work in New York, Milan and Hamburg, as well as throughout the UK. 95

Masayo Ave was born in Tokyo in 1962 but now lives in Milan. She studied architecture at Hosei University in Japan but after practising for some time, moved to Milan and graduated with an MA in Industrial Design from the Domus Academy in 1990. In 1992 she established Ave Design Corporation, based in Tokyo and Milan, and has won various awards in architecture and design. Clients have included international companies such as Authentics while her own design collection was launched by Atrox GmbH in 2000. 162, 163

François Azambourg was born in France in 1963 and studied Industrial Design in Paris. He has designed furniture and interiors for various Parisian shops, has produced work for VIA and numbers Cappellini, Plank, O Luce and Chainey among his most recent clients. He has also been involved in the development of new materials for Mandarina Duck, Hermès and Vuitton. In 1999 he was named Young Creator of the Year. 29

Shin and Tomoko Azumi studied industrial design at Kyoto City University of Art and the Royal College of Art, London. In 1996 they were finalists in the *Blueprint*/100% Design Awards. They founded Azumi's in 1995, undertaking projects for British, Italian and Japanese clients. They have exhibited at Sotheby's and at the Crafts Council, London. They are also frequent exhibitors at both the Cologne and Milan Furniture Fairs. 28

Jacob de Baan graduated from the Gerrit Rietveld Academy of Art in 1987 and worked in Amsterdam until 1991, when he moved to Germany to work for Team Buchin Design. In 1995 he moved back to Amsterdam and started the D4 Agency whose clients include the Dutch Ministry of Finance, Jumbo, NOVEM, Osram and Philips. 145

Emmanuel Babled was born in France in 1967, and today lives and works in both Milan and Venice. He studied at the Istituto Europeo di Disegno in Milan, then worked for several years with Prospero Rasulo and Gianni Veneziano at Studio Oxido in Milan. He has been working as a freelance designer since the early 1990s for clients such as Fine Factory, Steel, Casprini, Wedgwood, Waterford Crystal Ltd, Kundalini and Rosenthal. Since 1995 he has also been an independent producer of Murano glass vases and ceramic items. Using different kilns and collaborating with the master glass-blower, Livio Serena he has produced unique pieces for IDEE Co. Ltd, Tokyo. 153

Matthias Bader was born in 1970, trained as a toolmaker and then studied interior design at the Fachhochschule in Kaiserlautern, graduating in 1997. He worked for Hartz and Hermes and planned showrooms for Villeroy and Bosch worldwide. In 1999 he set up his own office and in 2000 joined forces with Andrea Winkler in Karshule, specializing in shop planning and design. Current clients include Pallucco and Phos Design. They have recently exhibited in Cologne and Milan. 123

Valter Bahcivanji was born in Brazil in 1958 and studied Industrial Design. He has designed furniture, household and technological goods and has been involved in product development for the domestic and office markets. His work is represented in the Museu da Casa Brasileira, São Paulo, Brazil. 103

Gijs Bakker was born in 1942 in The Netherlands. He studied Industrial Design in his home country and in Sweden. In 1966 he set up a studio in Utrecht with his wife, thereafter working as a freelance designer for clients such as Polaroid and Rosenthal. He has been Professor at the Design Academy, Eindhoven from 1987 and in 1993 he founded Droog Design with Renny Ramakers. Since then they have exhibited yearly in Milan and recent clients have included Mandarina Duck, Flos, Salviati and Levi's. 133

Philip Baldwin see Monica Guggisberg

Ralph Ball studied furniture design at Leeds College of Art and received a MDes from the Royal College of Art, London, where he now has been appointed Senior Tutor. He taught at various design schools through the UK and between 1981 and 1984 was a designer with Foster Associates. He is now a partner in his own firm Naylor-Ball Design Partnership. Ball's works can be found at the Contemporary Applied Arts and Crafts Council, the Victoria and Albert Museum, London and MoMA, New York. He was one of only seven European designers to be selected by Nobou Nakamura to develop furniture concepts for Japan. 94

Barber Osgerby Associates was founded in 1996 by Edward Barber and Jay Osgerby who met when they were studying together at the Royal College of Art, London. The practice has designed interiors for residential and commercial premises and has had furniture manufactured by Cappellini and the Conran shops. Its latest project is the Soho Brewing Company – a microbrewery and restaurant in London. The two were awarded Best New Designers in 1998 at the International Contemporary Furniture Fair in New York. Recent Projects include a pharmacy and herbal apothecary interior and a flagship hair salon for Trevor Sorbie. 38

Martine Bedin was born in Bordeaux, France, in 1957 and studied first at the Ecole d'Architecture de Paris, then in Florence. In 1978 she joined the Radical Design group and briefly collaborated with Ettore Sottsass before participating in the founding of the Memphis Group in 1981. Since 1982 she has divided her time between her design/architecture studio in Milan and Paris, where she teaches. In design her focus is furniture and lighting but since 1988 she has also worked on many architectural projects in France. She has taught and exhibited internationally and is the recipient of many awards. 80

Giampaolo Bendini was born in 1945 and studied architecture at Milan Polytechnic, after which he worked in architecture and industrial design. He was involved with public housing and restoration but also commercial and industrial sites. In product design he is particularly interested in the office and bathroom sectors and has been responsible for style at Bugatti and Lotus. Since 1973 he has been Agape's art director. Under his directorship, five Agape items have won the Design Plus Award. 218

Claudio Bellini was born in Milan in 1963. He graduated in Architecture and Industrial Design from Milan Polytechnic in 1990 and worked at Mario Bellini Associates until 1996, working on a diverse range of projects. In 1997 he founded Atelier Bellini, a new consultancy devoted to industrial design. His recent clients include Vitra, Heller, Artemide, Fiat, Venini, Driade, Rosenthal and Iguzzini and Fritz Hansen. 46, 88

Mario Bellini was born in 1935 and graduated in architecture from Milan Polytechnic in 1959. He began to design products and furniture in 1963 and has collaborated with numerous Italian and international manufacturers. Completed architectural projects include the Milan Trade Fair Extension; the Tokyo Design Centre; the Schmidtbank Headquarters in Germany; and the Natuzzi Americas Inc. Headquarters in North Carolina. As well as for architectural work, Bellini is well known for his exhibition design. He has lectured at leading design schools worldwide and since 1995 has taught in the School of Architecture at the University of Genoa. He was the editor of *Domus* from 1986 to 1991. Examples of his products can be seen in most major design collections including MoMA, New York. Bellini has received many awards including seven Compasso d'Oro prizes. 32

Sebastian Bergne studied industrial design at London's Central School of Art and Design and Royal College of Art, and worked in Hong Kong and Milan before forming his own practice, 'Bergne: Design for Manufacture' in 1990. His clients include Cassina, Vitra, Oluce, Authentics and Driade. He is a visiting tutor at Central Saint Martin's College of Art and Design and the London Institute as well as lecturer at the Royal College of Art. He was a jury member for the Design Week Awards in 1991 and has been a frequent guest on jury panels worldwide. He has taken part in numerous group shows in London, New York, Hamburg, Tokyo and Brussels, most recently 'Mutant Materials in Contemporary Design' at MoMA, New York in 1995 and in 1997 a one-man show at the International Contemporary Furniture Fair in New York. 130

Lars Bergström was born in 1962 in Stockholm and studied at the Royal Academy of Fine Arts. Since 1986 he has worked in collaboration with Mats Bigert (b.1965) and they have exhibited widely throughout the world. 161

Jeffrey Bernett founded Studio B in New York in 1995. In 1996 he presented an award-winning range of furniture accessories and objects at the New York Furniture Fair. He works in many areas, including interior architecture, furniture, lighting, graphic design and corporate image. His client list includes Authentics, B&B Italia, Cappellini, Dune, Hidden/sdb, Ligne Roset and Troy. He also lectures at schools in the US. 61, 168

Thomas Bernstrand was born in Stockholm in 1965. He studied at the University College of Arts, Crafts and Design, the Danish Designskole and the Inchbald School of Design, London, graduating in 1989. His work is frequently exhibited throughout Scandinavia and in London. In 1998 he won the Young Swedish Design Award. 85

Fabrice Berreux was born in France in 1964. After graduation from the Ecole Nationale des Beaux Artes in 1986 he worked in interior architecture, founding the group 18 Aout in 1987. In 1991, 18 Aout was reformed and enlarged into dix heures dix who have exhibited with VIA and whose designs are held in the collection of the Musée des Arts Décoratifs, Paris. Fabrice Berreux has taught throughout France and has recently been collaborating with other designers on various lighting projects. 95, 110

Marc Berthier is an internationally known industrial designer. He has received the Premio Compasso d'Oro in 1991 and 1994; the Form design award in Germany in 1995, 1997 and 1998; the Design Plus in 1999 and has also been awarded the Grand Prix National de la Creation Industrielle by the French Ministry of Culture. 180

Dinie Besems was born in 1966 and studied at the Gerrit Rietveld Academy, Amsterdam, graduating in 1992. Besems has exhibited often in The Netherlands and has won several awards for applied arts. 85

Jurgen Bey was born in The Netherlands in 1965 and studied at the Design Academy, Eindhoven where latterly he has taught. As an independent designer he has worked for Levi's and the *New York Times Magazine* and has participated in exhibitions in New York, Germany, Paris, and The Netherlands. 84, 85

Francesco Binfaré was born in 1939 in Milan where he still lives and works. From 1969 to 1976 he was a director at Centre Cassina. In 1980 he set up the Centre for Design and Communication for the realization of works such as Wink by Toshiyuki Kita, Feltri by Gaetano Pesce, the Collezione Il Cileo for Venini and designs for De Padova by Vico Magistretti. He has taught at the Domus Academy, Milan and the Royal College of Art, London, and held seminars and conferences in major institutions worldwide. His own client list includes Adele C, Cassina and Edramezzei. 45

Riccardo Blumer was born in Italy in 1959 and graduated in architecture from Milan Polytechnic. He is a practising architect and also designs furniture. His Lalegerra chair, designed in collaboration with Alias, was awarded the Compasso d'Oro in 1998. 26

Marc Boase was born in 1973 and graduated in 3D Design for Production from the University of Brighton, England, in 1997. He specializes in product, furniture and interior design and in 2000 was a finalist in the Peugeot Design Awards. 142

Cini Boeri graduated from Milan Polytechnic in 1951 and for some years collaborated with Marco Zanuso. In 1963 she set up her own studio concentrating on civil and interior architecture and industrial design. Her architectural projects in Italy and abroad include apartments, houses, shops, showrooms and office buildings. Her projects are represented in museums worldwide, and she has won many awards, including the Compasso d'Oro twice. 48

Jörg Boner is a product designer and member of the Design Group N2 based in Lucerne, Switzerland. He has exhibited in Hamburg, Cologne and Lucerne, and collaborated with Christian Deuber on a project for Dorbracht Interiors. He has also designed products for Hidden/sdb and is currently engaged in work for Hidden. 34, 64

Pierre Bouguennec was born in Brittany and moved to New York in 1987 where he trained as a cabinetmaker before starting practice as an architect. From 1989 to the present he has worked in his own studio, Boum Design, where he has carried out interior design, furniture and lighting projects. In 1998 his lamp, 'Plug In' was manufactured by Ligne Roset in France and won two awards at the Furniture Fair in Paris. Bouguennec is also a member of Samba Inc. He is currently working on a series of modular modern spaces. 58–9

Erwan Bouroullec was born in Quimper, France, in 1976 and studied industrial design at the Ecole Nationale Supérieure des Arts Appliques and the Ecole Nationale des Arts Decoratifs. Since 1998 he has been collaborating with his brother Ronan. 38, 41, 45, 145

Ronan Bouroullec was born in Quimper, France, in 1971. He graduated in applied and decorative art and has worked on a freelance basis since 1995, designing objects and furniture for Cappellini, Liagre, Domeau et peres, Ex Novo, Ligne Roset and Galerie Neotu. He was awarded Best Designer at the International Contemporary Fair, New York in 1999. 145

Box Design AB was founded in Stockholm in 1986 by Ann Morsing and and Beban Nord. Ann Morsing studied in San Francisco and the National School of Art & Craft, Stockholm, after which she worked for IKEA and Matell Arkitekter AB. Beban Nord studied art history and woodmanship in Stockholm. Before co-founding Box Design he worked in the Joinery at the Royal Palace, Stockholm and Svenska Rum Arkitekter. They have exhibited many times in Italy and Scandinavia. 200

Todd Bracher studied industrial design at St John's University, New York and furniture design at the Danish Design School, Copenhagen, graduating in 2000. He has acted as a consultant for various American and Italian companies and was Senior Designer for Bernstein Design Associates, New York from 1996 to 1998. Since 1998 he has been consultant designer for Posman Collegiate Bookstores Inc., New York. 24

Andrea Branzi was born in 1938 in Florence. He studied architecture, then founded the avant-garde group Archizoom Associates together with Gilberto Corretti, Paolo Deganello and Massimo Morozzi in 1966. From 1974 to 1976 he was involved with Global Tools, and in the late 1970s set up CDM, a Milan-based group of design consultants. He worked with Studio Alchimia and Memphis, designing furniture and objects and preparing shows and publications. He founded the Domus Academy in 1983 and has been its cultural director and vice-president. He teaches and holds conferences at universities in Italy and abroad, and has held many one-man shows at the Milan Triennale and at galleries worldwide. In 1987 Branzi won the Compasso d'Oro prize. 102, 122, 138, 139

Sergio Brioschi was born in Italy in 1960 and studied design at Milan Polytechnic. From 1981 to 1984 he worked for the architect and industrial designer Mario Bellini. In 1985 he began working with the designer Antonio Citterio and has designed for many prestigious clients including B&B Italia, Arc Linea, Olivetti, Vitra and Antonangeli. 104

Britefuture, led by Bastiaan Arler, is an Italian design team which creates sculptural objects, concerned with mechanics, assembly, purity of material and form. Dutch-born Bastiaan Arler has also lived and worked in Japan, Italy and Sweden. Before establishing Britefuture, he worked as a freelance designer for such clients as IKEA, Iittala Glass, Whirlpool, and on behalf of Studio de Lucchi for Olivetti and Fiat. 134

Bernard Brousse was born in Bordeaux, France, in 1948. From 1970 to 1980 he worked as a deep sea diver for French, Brazilian, British and Norwegian companies. In the early 1980s he started working in wood and design. Since 1989 has run a workshop where he designs and makes prototypes of lamps and furniture for different French and Italian companies. 105

Büro für form was founded in 1998 by the German designers Constantin Wortman and Benjamin Hopf. They work in the fields of interior and product design, particularly lighting and furniture. Objects play with the perception of the viewer and aim to blend organic shape with geometric elements. In 2000 they exhibited for the first time in Milan and Cologne. 104

Maximillian Burton was born in London in 1966. He studied Industrial Design at Manchester Metropolitan University and the Royal College of Art, London, graduating in 1992. After working for Pankhurst Design in London, the work from his Master's show was included in an exhibition at MoMA, New York and at the 'One Hundred Years at the RCA' show. He worked briefly for Hollington Associates, London, before joining Smart Design in 1995, where he is a senior designer. Since then he has worked on various independent projects which span both product and furniture design. 218

Sigi Bussinger was born in 1965 in Munich. He studied wood carving and since 1989 has been working as a sculptor and performing artist. 94

Bute Fabrics was founded nearly fifty years ago on the Isle of Bute, Scotland, and it is among the company leaders in the industry. Fabrics fall into three groups: project driven, off the peg and bespoke, such as the ten miles of fabric ordered for Chek Lap Kok Airport, Hong Kong. Specializing in wool rich blends, Bute upholstery fabrics are used in auditoria, corporate interiors, hotels and restaurants, airports and lounges, banks and governmental buildings. 171

Louise Campbell graduated from the London College of Furniture in 1992 and completed further study at the Danish Design School's Institute of Industrial design in 1995. The following year she set up her own studio in Copenhagen and has since been involved in various projects including setting up the 'Walk the Plank' project with Sebastian Holmback and Cecilia Enevoldsen. 22, 74, 75

Massimo Canalli studied design in Milan and then moved to New York to work with Wajskol Inc. He currently works as a graphic designer in Michele De Lucchi's studio but also has his own Milan based studio where he works on corporate image, packaging, editorial design and web design. 187

Chiara Cantono studied architecture and industrial design at Milan Polytechnic. She formed her own design studio in 1993, her work being based on experimentation with and study of new materials and technologies. In 1997 she created a new collection using material which up until then had been used solely in Formula 1 car racing and in the aeronautical industry. The following year Cantono presented 'Alicia' a collection of lamps which react to light like fibre optics. She has also collaborated with companies such as Brunati Italia and the Busnelli Industrial Group. 50

Marco Carenini was born in Zurich in 1968 and studied architecture and design. In 1996 he set up the studio B&C Design where he is working independently in furniture, product, graphic and corporate design. 91, 102, 123

Caroline Casey studied fashion and textile design at Sydney College of the Arts, graduating in 1986. Since then she has worked for may companies in New York, London and Sydney as a fashion, interior and furniture designer. In 1999 she had a solo exhibition at the Powerhouse Museum, Sydney. 18

Jean-Charles de Castelbajac is a Parisian couturier. Born in 1949, he has been creating garments since the age of 19, searching for a pure, elemental, visual language. In 1979 he began diversifying into other areas of design and has recently worked with Ligne Roset. He has taught at Central Saint Martin's School of Design, London and the Academy of Applied Arts, Vienna. 168

David Chipperfield was born in London in 1953 and graduated from the Architectural Association in 1977. He became a member of the Royal Institute of British Architects in 1982 and after collaboration with Douglas Stephen, Richard Rogers and Norman Foster, founded David Chipperfield Architects in 1984. The studio's activities are varied, ranging from furniture design to interior architecture and urban planning. In 1987 he opened an office in Tokyo, followed by ones in Berlin and New York. He has since been able to undertake projects for Issey Miyake, the Neues Museum, Berlin and the Hotel Shore Club, Miami. Recent projects include plans for the enlargement of the San Michele cemetery in Venice and the Museum of Art, Davenport USA. He has won many awards and also been Professor of Architecture at the Staatliche Akademie, Stuttgart and visiting professor at various institutions. 54

Aldo Cibic was born in Schio, Italy, in 1955 and moved to Milan in 1979 to work with Ettore Sottsass, participating in the foundation of the Memphis group. In 1989 he set up his own independent design practice, Cibic & Partners, a studio whose interior design and architecture projects range from department stores to cinemas, public works to private residences. 150

Antonio Citterio was born in Meda, Italy, in 1950 and has been involved in industrial and furniture design since 1967. He studied at Milan Polytechnic and in 1973 set up a studio with Paolo Nava. They have worked for B&B Italia and Flexform, among others. In 1979 they were awarded the Compasso d'Oro. In 1987 Terry Dwan became a partner in Studio Citterio Dwan, and interior design projects have included schemes for Esprit and offices and showrooms for Vitra. Among work in Japan, in partnership with Toshiyuki Kita, are the headquarters in Kobe for World Company, the Corente Building in Tokyo, and the Diago headquarters in Tokyo. Citterio has taught at the Domus Academy in Milan and participated in many exhibitions including independent shows. In 1993 he designed the exhibition 'Antonio Citterio and Terry Dwan' promoted by Arc en Rêve in Bordeaux, which travelled to Osaka and Tokyo in 1994. In 1996 Antonio Citterio participated in numerous design competitions including the corporate design for the Commerzbank pilot branches in Germany, the new retail environment for Habitat in Paris and the restructuring of the existing Line 1 metro system of Milan. Recent projects include interior architecture for Cerruti, Emanuel Ungaro SA, the UEFA headquarters, B&B Italia and Zegna Sports clothing store. In 1999 Antonio Citterio & Partners was established as a multidisciplinary architecture and design studio with offices in Milan and Hamburg. 47, 62, 72, 120

Toni Cordero was born in Italy in 1937. After studying architecture in Turin he opened a design office in 1962. Architectural clients have included Olivetti, Fiat, Blumarine, Kenzo, the Banco Mediceo del Filarete, and the Turin Automobile Museum. His furniture and design work includes collaborations with Acerbis, Driade, Artemide and Sawaya & Moroni. 106

Carlo Cumini was born in Udine in 1953 where he lives and works. He studied at the Marinoni Technical School and started working for the family firm. When Horm was established he became the main furniture designer, concentrating on design in solid wood. Since 1991 he has also worked for Luc-e and Nuova Auras. He has collaborated with Alberto Freschi and Carlo Biancolini; he continues to work in the design field, investigating new materials. 81

dai design and identification was founded in Zurich in 1987 as a design agency, specializing in architecture, interior design and corporate identity. 208

Lorenzo Damiani was born in 1972. He studied architecture at Milan Polytechnic and since then has taken part in many exhibitions and competitions. In 1999 he represented Milan in the Design section at the Biennale of Young Artists and the Meditteranean. 73

Christopher C. Deam, principal of San Francisco-based CCD Architecture and Furniture Design, is a Milan-trained architect whose furniture designs have generated many awards and international recognition. He received the Editor's Award for Furniture Design in 1997 and his work is in the permanent collection of the San Francisco Museum of Modern Art; it has also been exhibited widely. 225

Design 3 Produktdesign was founded in Hanover in 1987 as the German office of Moggridge Associates (London) and ID Two (San Francisco). In 1990 it became an independent company when the others joined IDEO. in 1999 the studio moved to Hamburg. 186

Christian Deuber is one of the founder members of the N2 group in Lucerne, Switzerland. While training as an electrical engineer, he gained experience in lighting design and set up his own studio, Pharus Lighting Design. He has designed lighting for Driade, Palluccoitalia and bathrooms for Dornbracht Interiors with Jorg Boner. Most recently he has designed for Hidden/sdb. 64

Jane Dillon was born in the UK in 1943 and studied interior design at Manchester College of Art and Design followed by an MA at the Royal College of Art, London. Since graduation in 1968 she has developed an international client list which includes Olivetti, Casas Barcelona, Habitat, Heals, Thonet, Cassina, Herman Miller and Ercol. In 1972 she formed a design partnership with Charles Dillon – Studio Dillon has had a range of own brand products since 1997. Recent projects have included seating design for the Science Museum, London, consultancy work for the Globe Theatre, London and a collection for W. Lusty and Sons. She teaches at the Royal College of Art and is currently involved in projects with Guy Mallinson Furniture Ltd and Keen Group Ltd. 19

Felice Dittli studied design in Basel, Switzerland specializing in interior decoration, products and construction design. In 1988 she set up her own studio and now works in interior decoration and product design. She has had various exhibitions and won several awards. 117

Dante Donegani and Giovanni Lauda established their design company in 1993. Donegani graduated from the Faculty of Architecture in Florence after which he worked for Corporate Identity Olivetti. Since 1993 he was been the Director of the Design Master Course held at the Domus Academy. He has collaborated with firms such as Memphis, Stildomus and Luceplan and has won several major awards for his architectural schemes such as the Manhattan Waterfront, New York (1988) and the Berlin Wall Prize (1987). Lauda has a degree in architecture and works in the fields of interior, exhibition and industrial design for companies such as Artas, Play Line and Sedie and Company. He was a member of Morozzi and Partners until 1994 when he became responsible for the Design Culture course in the Industrial Design Master Course at the Domus Academy. He has curated numerous exhibitions including 'Il Design Italiano dal 1964–1990' which was held at the Triennale of Milan in 1996. 68, 69

Rodolfo Dordoni was born in Milan in 1954, where he studied architecture. He is involved in art direction, corporate identity, interior design, and furniture and lighting design. His client list includes Acerbis, Arteluce, Artemide, Cappellini, Dolce & Gabbana, Driade, Ferlea, Flos, Molteni, Moroso and Venini. 56

Dumoffice is based in Amsterdam and consists of the designers Wiebe Boonstra, Martijn Hoogendijk, and Marc van Nederpelt. All three graduated in 1994 from the Design Academy Eindhoven and have produced work for Belux of Switzerland and sdb Industries, The Netherlands. They have exhibited in Milan and are also represented in the Stedelijk Museum, Amsterdam and the San Francisco Museum of Modern Art. Their aim is to produce work which is unorthodox with an element of wit. 89

Knuth Eckhard was born in Hamburg in 1945 and studied art history and theatre. While working as a dramatic advisor, he found that he preferred working as a light designer. He first employed laser technology in 1979 to 1980 and attended the School of Holography in San Francisco. In Munich he built a machine for 360° holography and his hologram of the crown of the Holy Roman Emperors was the focal point for the Millennium Exhibition in Austria. In 1997 he began a collaboration with Ingo Maurer and developed the Holonzki holographic lamp. 96

Claus Christian Eckhardt was born in 1965 and studied industrial design at the Academy of Fine Arts, Brunswick, going on to work as an interior designer for Silvestrin Design in Munich. From 1992 to 1999 he was in charge of designing consumer electronics and communication products for the Blauplunkt design department in Hildesheim, where he was also responsible for the design of Bosch mobile phones. He is now Chief Designer and Head of Global Product Design for Bosch Telecom in Frankfurt. He has received several international awards. 186

Cecilia Enevoldsen was born in 1970 and graduated in Industrial Design from Denmark's Design School in 1995. In the same year she established her own design company which mainly designs furniture. She exhibits often and has won several design prizes. Along with Sebastian Holmback and Louise Campbell she initiated the 'Walk the Plank' project. She lectures at the Danish Design School and has worked as an adviser at the Architecture School, Copenhagen. 22

David Farrage was born in England and studied Industrial Design at the Royal College of Art, London. He worked for Ross Lovegrove Studio X, moved to Japan to work for the architectural and urban product manufacturer Sekisui Jushi, and moved back to London to work for FM Design. In 1996 he became Senior Industrial Designer at the New York-based Smart Design. More recently he has joined Sony at its US East Coast Design Centre. 210

Khodi Feiz was born in Iran in 1963. He graduated from Syracuse University, USA in 1986, and worked for Texas Instruments Design Centre, joining Philips Design in 1990. He became Manager and Senior Designer for the Advanced Design Group but left to establish Feiz Design Studio in 1998. Based in Amsterdam, Feiz Design Studio has been featured in exhibitions worldwide and its current client list includes Alessi, LEGO, Loewe, LG Electronics and Polydor. 44

Norman Foster was born in Manchester, England in 1935 and studied at the University of Manchester and Yale University. He established Team 4 in 1963 – with his late wife, Wendy, and Su and Richard Rogers – and founded Foster Associates in 1967. He is renowned for his high-tech designs, such as the Hongkong and Shanghai Bank (1979–85) and Stansted Airport (1981–89). More recent projects include the Sackler Galleries at the Royal Academy of Arts, London; the Centre d'Art Cultural, Nimes; the new headquarters for Commerzbank in Frankfurt, the remodelling of the Reichstag in Berlin and the vast new airport at Chek Lap Kok for Hong Kong. He is currently working on the new Wembley Stadium in London. His work has won over 60 awards and citations. Although primarily concerned with large-scale architectural projects, Foster is also active in furniture and product design. 53

Nobuhiro Fujii was born in 1968 in Japan. He attended Tsukuba University and joined Sharp Corporation in 1992 as a product designer, working on home appliances. 198

Naoto Fukasawa was born in Kofu, Japan, in 1956 and graduated from Tama Art University. He was Chief Designer at the Seiko Corporation before joining IDEO San Francisco. He returned to Japan in 1996 to become Director of IDEO, Japan. He has won many awards and his Metro stacking chair is included in the collection of the San Francisco Museum of Modern Art. His work was recently exhibited in Berlin as part of the exhibition 'Where We Stand'. He has lectured at the Royal College of Art, London, at Tama Art University and is currently retained as teacher to Matsushita Design. 202–3

Jorge Garcia Garay was born in Buenos Aires, Argentina, and has worked in Barcelona since 1979 as the director of Garcia Garay Design. He is involved almost exclusively in lighting with an architectural design emphasis and his work can be seen in permanent collections in Europe and the United States. His work has been published in many journals, including *Blueprint*, *Abitare* and *Interni*. His products are distributed worldwide. 111, 112

Kristian Gavoille was born in Brazzaville in 1956 and has lived in Paris since 1986. From 1986 to 1991, after practising as an architect with DPLG Toulouse, he collaborated with Philippe Starck on several hotel and restaurant projects and in 1992 was voted designer of the year. He has designed shops for Kookai and furniture clients include Ligne Roset and Mobilier National. 169

Bruno Gecchelin was born in 1939 and studied architecture at Milan Polytechnic. Since 1962 he has been involved in many areas of industrial design numbering such companies as Indesit, Olivetti, Arteluce-Flos, Fiat, Antonangeli, Venini, Matsushita, Salviati and Tronconi as clients. He has won the Compasso d'Oro prize in 1989 and 1991. 117

Massimo Iosa Ghini was born in Bologna in 1958 and studied architecture at Milan Polytechnic and in Florence. In 1985 he began working for RAI. His work with Memphis dates from 1986, the year he founded the Bolidismo movement. In industrial design he has collaborated with Cassina, Flou, Mandarina Duck, Alessi, Akfi, Ritzenhoff, WMF, Yamigawa and Hasashi Glas. 200

Christian Ghion was born in Montmorency, France in 1958. He graduated from Etude et Creation de Mobilier, Paris in 1987 and since then has worked with Patrick Nadeau. In 1998 he started his own project concentrating on industrial and interior design for European and Japanese companies including Cinna/Roset, Neotu, 3 Suisses, Idee, Tendo and Thierry Mugler. His work has been awarded several prizes and is on show in the design collection of the major museums of New York, Los Angeles and Paris. 43, 151

Stefano Giovannoni was born in La Spezia, Italy, in 1954 and graduated from the architecture department at the University of Florence in 1978. From 1978 to 1990 he lectured and carried out research at Florence University and also taught at the Domus Academy in Milan and at the Institute of Design in Reggio Emilia. He is the founding member of King-Kong Production, which is concerned with avant-garde research in design, interiors, fashion and architecture. Clients include Alessi, Cappellini, Arredaesse and Tisca France. In 1991 he designed the Italian Pavilion at 'Les Capitales Européennes de Nouveau Design' exhibition, which was held at the Centre Georges Pompidou in Paris. 200, 219

Ernesto Gismondi was born in 1931 in San Remo, Italy. He studied at Milan Polytechnic and the Higher School of Engineering, Rome. In 1959, with Sergio Mazza, he founded Artemide SpA, of which he is the president and managing director. Since 1970 he has designed various lights for Artemide and in 1981 he was involved in the development of the influential Memphis group. He has sat on the boards of several national design boards and taught widely in the same field. 116

Marco Giunta was born in 1966 and studied design and architecture at the Polytechnic of Milan. From 1992 to 1996 he acted as a consultant for Zerodisegno and Quattrocchio. Since 1994 he has taught at Milan Polytechnic and also in Antwerp and Genova. In 1995 he set up Disegni which specializes in furniture, displays for shops and packaging. 83

Natanel Gluska was born in Israel in 1957 but currently works n Zurich and in Lefkada, Greece. He studied in Israel, the Hague and the Rietveld Academy, Amsterdam. His most recent projects include the interior of a club in Zurich, chairs for the Union Bank of Switzerland (both 1999), chairs for the Sanderson Hotel, London, and Donna Karan's shop, Madison Avenue, New York (both 2000). 14, 21

Anki Gneib is an architect and designer based in Stockholm. Born in London in 1965, she studied at the University College of Arts, Crafts and Design in Stockholm and at Middlesex Polytechnic, UK. Since 1993 she has worked independently but has also undertaken public and private interior design jobs through Wallenstreen and Ostgren. Clients include Asplund, Fogia, Arvesund and Interstop. She has exhibited throughout Europe and Japan and won several awards. 161

Tobias Grau was born in Hamburg. He studied economics in Munich and then design in New York, thereafter working in the product development department of Knoll International. In 1984 he began work as a freelance interior designer, and since 1985 has developed an internationally recognized and exclusive lighting range. In 1999 the first two Tobias Grau shops were opened in Hamburg and Berlin, the third in Dusseldorf in 2000. 90, 124

Johanna Grawunder is an architect based in Milan, where she has worked with Sottsass Associati since 1985, becoming a partner in 1989. Born in San Diego in 1961, she studied architecture in California, San Luis and Florence. She has collaborated with many companies including Egizia, WMF and Giotto as well as designing one-off pieces for the Venetian company Salviati and Christophile of France. 112

Konstantin Grcic was born in 1965 in Germany. He trained as a cabinetmaker and continued his education at the John Makepeace School for Craftsmen, then studied design at the Royal College of Art, London, on a scholarship from Cassina. He worked in the studio of Jasper Morrison in 1990 and founded his own studio in Munich. Clients have included the Munich Tourist Office (in collaboration with David Chipperfield Architects UK), Agape, Cappellini, ClassiCon, Flos, littala, Montina and Zeritalia while recent commissions have comprised work for Authentics GmbH and Whirlpool Europe srl. 26, 120

Group Kombinat is Sven Anwar Bibi, Mark Gutjahr and Jorg Zimmerman. They all studied design at the University of Applied Sciences, Cologne and frequently exhibited their work in both Cologne and Milan. In 2000 Group Kombinat split up but Mark Gutjahr and Sven Anwar Bibi are continuing to work together under the name Bibi*Gutjahr. 34

Gitta Gschwendtner was born in 1972 in Wurzberg, Germany. She holds at BA (Hons) in Furniture and Product Design from Kingston University, London, and an MA in Furniture Design from the Royal College of Art, London. Since 1998 she has worked as a freelance furniture, product and interior designer. 19

Dögg Gudmundsdóttir was born in Reykjavik, Iceland and holds a Diploma in Industrial Design from the Istituto Europeo di Design in Milan. Gudmundsdóttir has held many exhibitions within Scandinavia and currently works for IKEA as a freelance designer. 24

Monica Guggisberg and Philip Baldwin are an American/Swiss couple who have been working together for 15 years. They trained in Sweden at the Orrefors Glass School and later in the studio of Ann Wolff and Wilke Adolfsson. In 1982 they established their own design and hot glass studio near Lausanne in Switzerland. Their creations have been exhibited widely in Europe, America and Japan and are to be found in many museums worldwide. 144

Marti Guixé was born in 1964 and studied Interior Design in Barcelona, and Industrial Design in Milan. From 1994 to 1996 he was a consultant for KIDP, Seoul. Recent projects include shops for Camper in Barcelona, London and Via Montenapoleoni, and an apron and bag collection for Authentics. He has exhibited with the Droog Design collective, individually in Barcelona, Milan and Berlin and currently works between Barcelona and Berlin as a Techno-gastrosof, Tapaist and Designer. In 1999 he won the Ciutat de Barcelona Design Prize. 84, 85

Alfredo Häberli was born in Buenos Aires, Argentina, in 1964 but moved to Switzerland in 1977. He studied industrial design in Zurich and worked as an installation designer at the Museum fur Gestaltung until 1993. Since then he has worked alone or in collaboration with Christophe Marchand for companies such as Alias, Edra, Zanotta, Thonet and Driade. He is currently developing products for Asplund, Bally, Cappellini, littala and Zeritalia. 38, 40, 49, 81, 122, 160, 161

Zaha Hadid was born in Baghdad in 1950 and studied at the American University in Beirut and the Architectural Association in London. She joined the Office for Metropolitan Architecture and worked on the Dutch Houses of Parliament Extension in The Hague. She established her own practice in 1979 and in 1983 won the Hong Kong Peak International Design Competition. Recent projects include the Contemporary Arts Centre in Cincinatti and an exhibition pavilion in Weil am Rhein, Germany. She has taught at the Harvard Graduate School, Illinois University and the Architectural Association, London. She currently teaches at the Hochschule in

Hamburg and at Columbia University, New York. 52

Makoto Hashikura was born in Japan in 1972. He joined Mitsubishi in 1998 after graduating from Keio University. Since then he has designed computers and visual information projects such as LCD displays and CRT Monitors. 207

Sam Hecht was born in London in 1969. After studying at the Royal College of Art he worked as an interior and industrial designer. He moved to Tel Aviv and joined the Studia group, then moved to San Francisco where he began collaboration with IDEO. He also completed projects for AT&T and NEC. He then worked for IDEO in Japan with clients such as NEC, Seiko and Matsushita. He won the D&AD Exhibition Category Award in 1998 for his White Box design, and moved to London to become Head of Industrial Design at IDEO. Recent projects include the Thames Water Pavilion at the Millennium Dome, Greenwich. He lectures in Japan and his work forms part of the permanent collection of MoMA, New York. 202–3

Steven Holl is the principal of Steven Holl Architects, New York. Born in 1947 in Bremerton, Washington, he attended the University of Washington, then studied architecture in Rome and at the Architectural Association, London. He established his firm in 1989 and his work was presented in a two man show at MoMA, New York. Recent awards include the 1996 Architecture Awards for Excellence in Design. Holl's Chapel of St Ignatius in Seattle won a National AIA Award for Excellence in Design in 1997. 110

Geoff Hollington was born in the UK in 1949. He studied industrial design at the Central School, London and environmental design at the Royal College of Art, London. In 1980 he established the Hollington studio which has such international clients as Kodak, Gillette, Ericsson and Herman Miller. His studio's products are in museum collections and have won many international awards. The studio is now a leader in the field of digital interaction design and is currently collaborating with the Science Museum, London. He is a fellow of the Royal Society of Arts, the Chartered Society of Designers and a member of the Industrial Design Society of America. 67, 181

Bohuslav (Boda) Horak was born in Pardubice in the Czech Republic in 1954 and attended both the Zizkov Art School of Prague and the Academy of Applied Arts. In 1987 he became a member of the design group Atika, also in Prague, and in 1990 opened his own studio. His designs are produced frequently by Anthology Quartett. 149

Isao Hosoe was born in Tokyo in 1942 and studied aerospace engineering at the Nihon University of Tokyo. Since 1967 he has lived in Milan and collaborated with Alberto Rosselli until 1974. He is Professor of Industrial Design at Milan Polytechnic. Among his many awards are the Compasso d'Oro and the Biennial of Industrial Design at Lubiana. He has exhibited many times and his works are in the permanent collections of the Victoria and Albert Museum, London; the Centre Georges Pompidou, Paris; the Chicago Athenaeum; and the Museum of Science and Technology, Milan. 92, 93

Elina Huotari graduated with an MA from the University of Art and Design, Helsinki in 1999. Since 1996 she has worked as a designer for Ratti SpA, Italy, Studio Lasse Keltto and Verso Design Oy, Finland. Her work has been exhibited in the UK, Italy and Finland. 174

Richard Hutten graduated in industrial design from the Design Academy in Eindhoven in 1991 and then set up his own studio. He has achieved international recognition for his innovative designs which can be seen in the permanent collections of many design museums in The Netherlands as well as in the Vitra Museum, Weil am Rhein. He has held exhibitions worldwide and Philippe Starck selected two of his products for the interiors of the Delano Hotel in Miami and the Mondrian Hotel in Los Angeles. Hutten taught product design at the Art School of Maastricht from 1996 to 1998. 34, 35

Takashi Ifuji was born in 1969 in Gifu, Japan. After studying industrial design at the University of Tsukuba he joined the Design Laboratory at Fujitsu Ltd. In 1999 he moved to Milan and established Ifuji Design. He took part in the Salone de Mobile in Milan and was awarded the Red Dot for the Highest Design Quality in 1999, Design Zentrum Nordrhein Westfalen, Germany. 145

Yoshinori Inukai was born in Nagoya, Japan in 1964. He graduated from Kanaza College of Art in 1996 and joined Canon Inc. shortly afterwards. He has received numerous awards and in 1999 was given the IF Award in Germany as well as the IDEA 99 bronze prize, USA. 193

James Irvine was born in London in 1958. He studied design at Kingston Polytechnic and then at the Royal College of Art, London, graduating in 1984. He moved to Milan where he was a design consultant for Olivetti design studio Milan, designing industrial products under the direction of Michele De Lucchi and Ettore Sottsass. In 1988 he worked for the Toshiba Design Centre in Tokyo but returned to Milan to open his own studio whose clients included Cappellini, BRF and SCP. From 1993 to 1998 he was also a partner in Sottsass Associati. His first personal exhibition was at the Royal College of Art, Stockholm in 1993. In 1999 he completed the design of the new city bus for the Hannover transport system, USTRA. Current clients include Artemide, B&B Italia, Magis, Whirlpool and Canon Inc., Japan. 118

Noriaki Itai was born in Japan in 1965 and joined Sharp Corporation in 1983. He has designed electronic component-based application projects and is now in charge of product design development for television, video and DVD. 192

Airi Itakura was born in Kanagawa, Japan in 1974. She graduated from the Women's College of Arts in 1996 and joined Canon Inc. shortly afterwards. She designs scanners and fax machines and has won the Japanese Good Design Award (1999). 193

Setsu Ito was born in Yamaguchi, Japan in 1964. He obtained a Masters Degree in product design from the University of Tsukuba and has since published studies on product semantics and design valuations for the Japanese Society for the Science of Design. He has undertaken design research projects for the TDK Corporation, NEC Electric Co and Nissan Motor Co, and in 1989 worked for Studio Alchimia in Milan. Since 1989 he has collaborated with Angelo Mangiarotti. He is currently a consultant designer for the TDK Corporation, Hitachi and Casio and since 1996 he has taught at the Instituto Europeo di Design in Milan. Some of his works are in the permanent collection of the Die Ne Sammlung Museum in Munich. 209

Dakota Jackson has been a considerable presence in American Furniture for three decades, with furniture in the permanent collections of the Cooper Hewitt National Design Museum, the Brooklyn Museum, the American Craft Museum, the Chicago Atheneum, the London Design Museum and te Deutsche Architekturmusuem, Frankfurt. Recently Jackson has completed designing the Steinway Tricentennial Grand Piano. 81

Hans Sandgren Jakobsen was born in 1963. He trained as a cabinetmaker before studying furniture design at the Danish Design School in Copenhagen. In 1991 he began working at Nanna Ditzel's drawing office but started his own practice in 1997. He is also a member of the design group 'Spring' and the Danish Designers MDD. Freelance clients include Via America, Fritz Hansen and Fredericia Stolefabrik. He has exhibited his work widely. 22, 23, 57

Jam Design was formed in 1994 by A. Zala and M. Paillard. Collaborations have included work for Philips, Sony and Zotefoams Plc. 1999 saw the launch of the Flatscreen coffee table and 2000 of Panel Light, a wall mounted lamp. One off projects have included the Millennium Crib. 212

Claus Jensen is an Industrial Designer. Born in 1966 he graduated from the Danish Design School in 1992 and established Tools Design with Henrik Holbaek. Holbaek was born in 1960 and graduated from the Royal Academy of Arts Architectural School in 1990. Since then they have exhibited widely throughout the world and won many awards. 131, 201

Carsten Joergensen was born in 1948 in Denmark. He was educated at the Art and Craft School in Copenhagen where he studied painting and graphic illustration. In 1974 he started his collaboration with Bodum working as a freelance designer and later founding a department within the company which he later moved to Lausanne. 156, 200

Claudy Jongstra was born in The Netherlands in 1962 and studied Fashion Design in Utrecht. She has designed fabrics for a diverse range of clients including John Galliano, Donna Karan, Volvo and 'Star Wars, the Phantom Menace'. Her work has been exhibited worldwide and she is represented in museums in London, New York, Belgium and The Netherlands. 164

Yehudit Katz was born in Israel and graduated from the Bezalel Academy of Art and Design, Jerusalem. She received a postgraduate degree from the Ateneum School of art in Helsinki and has since worked on dobbies and jacquards for various mills in Israel and abroad. She has taught weaving at the Art Centre of Jerusalem and is Senior Lecturer at the Shenkar College in Ramat Gan. She received the Alix De Rothschild Foundation Prize for Fibre Arts in 1998 and has now retired from industrial design in order to research the interaction between weaving and light. 172

Yaacov Kaufman was born in Russia in 1945, lived in Poland until 1957, and moved to Israel where he now lives and works. He studied at the Bat Yam Institute of Art and is now Professor of Industrial Design at the Bezalel Academy of Art and Design, Jerusalem. He is particularly active in the fields of furniture and lighting design and has won several international design prizes. He is currently collaborating with several Italian furniture designers and has designed for Lumina for many years. 124

Hideki Kawai was born in Nagoya, Japan in 1967. He graduated from the Aichi College of Art in 1989 and joined Canon Inc. as an industrial designer. In 1999 he won the Good Design Award in Japan. 181

Makoto Kawamoto was born in Japan in 1965 and graduated form Osaka's University of Arts. In 1994 he moved to Perugia, then Milan and in 1995 began a collaboration with Sawaya and Moroni. He is also active in graphic and interior design, having designed the offices for Il Cittadino in Milan in 1996. Kawamoto exhibits regularly, most recently with the Borderlight Group in Milan 2000. 93

King Miranda Associati was founded in 1976 by Perry King and Santiago Miranda. They work in product, service, interior and exhibition design and are active in a variety of industries from consumer products to furniture, lighting to telecommunications. With clients worldwide, they founded the European Designer's Network in 1990. 27, 123

Toshiyuki Kita was born in Osaka in 1942 and graduated in industrial design from the Naniwa College in 1964. He established his own design studio in Osaka and began working both in Milan and Japan focusing on domestic environments and interior design. In 1989 he was presented with the Delta de Oro Award in Spain. He designed the chairs and interior for the rotating theatre at the Japanese Pavilion at Expo '92, Seville. He is a visiting lecturer at the Hochschule für Angewandte Kunst in Vienna and has founded a private school in the Fukui Prefecture of Japan. 194

Henrik Kjellberg and Mattias Lindqvist are the names behind the award-winning design company Sweedish, which was founded in 1998. Henrik was born in 1971 and studied in Oslo, New York and Stockholm while Mattias was born in 1967 and studied in Stockholm and Copenhagen. Both graduated in furniture and interior architecture and their notable clients have included IKEA of Sweden. 94

Kazuyo Komoda was born in Tokyo in 1961 and studied at the Musashino University of Art. She began her career in industrial and interior design in 1982 and since 1989 has worked in Milan. After collaboration with Denis Santachiara she established her own studio in Milan. She works, among others, for Acerbis International, Bernini, Driade, Rosenthal and Yamaha. 210

Tobias Koeppe was born in Wolfsburg, Germany, in 1958 and trained as a metalworker. He studied industrial design in Hanover and after working in the industry for some years founded his own studio in 1996. He focuses on the creation of new product ideas and the design of high quality goods in metal, plastics, glass and porcelain. 205

Komplot Design was founded in 1987 by Poul Christiansen and Boris Berlin and is active in the fields of Industrial, furniture and graphic design. Poul Christiansen was born in Copenhagen in 1947, where he studied architecture at the Royal Academy of Fine Arts. Boris Berlin was born in Leningrad in 1953, where he studied at the Institute of Applied Arts and Design and moved to Denmark in 1983. They have exhibited widely in Europe and Asia and have won many awards in Scandinavia. Their work is in the permanent collections of the Danish National Art Foundation and the Danish Museum of Decorative Art. 22

Geert Koster was born in The Netherlands in 1961. He studied in Groningen and Milan, graduating in 1985. Since then he has been based in Italy, collaborating with Michele De Lucchi, participating in the Solid group and co-founding the ecological design group 'o2'. For Studio De Lucchi he worked with clients such as Vitra, ENEL produzione, Telecom Italia and Mandarina Duck. In 1989 he opened his own studio in Milan, working on interiors, exhibitions, furniture and industrial design for clients such as Abet Laminati, Olivetti, Cappellini and Hidden. In 1999 he co-founded Park Studio (architecture and design). 34, 35

Defne Koz was born in Ankara, Turkey and moved to Milan in 1989 to study for an MA in Industrial Design at the Domus Academy. She spent 1991 to 1992 at Sottsass Associati but since 1992 has undertaken independent work in interior planning and industrial design, working for Cappellini and Ala Rosa among others. Recently she has planned interiors for private and commercial customers and designed office furniture for a Turkish company. Her projects have been shown in Milan, Cologne and Istanbul. 64

Marc Krusin was born in London in 1973 and graduated in Furniture Design from Leeds Metropolitan University. Following placements in the UK and Italy, he moved to Milan and collaborated with various prestigious international clients including Piero Lissoni's office. In 1998 he co-founded the Milan-based design group Codice 31, which has continued to expand. His clients include Bosa, Fontana Arte and Saporiti. 101

Katsunori Kume was born in Japan in 1967. In 1989 he graduated in Industrial Design from Kuwazuma Design School and joined Sharp Corporation. He is responsible for product design development in the field of television, LCD projector, video camera and advance design development. 183

Tsutomu Kurokawa was born in Aichi, Japan, in 1962. He studied design at Tokyo Designer School in Nagoya and trained in the offices of Ics Inc. and Super Potato Co Ltd. He established H. Design Associates with Masamichi Katayama in 1992. They are involved in interior and furniture design and have worked on a series of boutiques in Japan. They have exhibited their work internationally. In 2000 Kurokawa established Out DeSign. 107

Vardit Laor was born in 1972 in Rehevot, Israel. She graduated from the Bezalel Academy of Art and Design, Jerusalem, in 1998 and exhibited in Milan in 2000. Her focus is on industrial, furniture and exhibition design. 29

Danny Lane was born in Urbana, Illinois in 1955. Largely self-taught, he moved to England in 1975 to work with the stained-glass artist Patrick Reyntiens, then studied painting at Central Saint Martin's College of Art and Design in London. In 1983 he co-founded Glassworks with John Creighton and began a three-year association with Ron Arad. He has worked with metal and wood and participated in numerous museum and gallery exhibitions and international furniture shows. In 1988 he held three one-man shows in Milan, London and Paris and started producing work for Fiam Italia. In 1994 he was commissioned by the Victoria and Albert Museum in London to install a balustrade of stacked glass in the Glass Gallery. In 1998 he created a water sculpture for the Conrad International Hotel in Cairo and also developed the technique of blowing borosilicate glass into tubular shapes, creating a glass fountain in Shanghai. In 1999, Lane held an exhibition in London, 'Breaking Tradition', and in the past year has designed glass-based pieces for hotels in Singapore, Hong Kong and the ITN Building, London. 30, 153

Taco Langius was born in The Netherlands in 1964 and studied in Eindhoven, then Milan. Since 1990 he has collaborated with Philips Whirlpool, Matteo Thun, Aldo Rossi

Architecture and Luca Trazzi Industrial Design and Architecture. Currently he works with Lissoni Associates and is also a member of Codice 31, Milan. 102

Kristiina Lassus was born in 1966 in Helsinki, Finland. She studied interior and furniture design at the University of Industrial Arts, Helsinki and then at the National College of Arts, Crafts and Design in Stockholm. She has worked in Helsinki and Brisbane, specializing in interior and graphic design. In 1993 she set up the D'Imagio design practice and has been working with Zanotta since 1995. 138

Marta Laudani and Marco Romanelli work together as interior and product designers. Laudani graduated in Rome in 1979. Her competition project for the piazza in the Parco dei Caduti in Rome won first prize and executive status in 1990. Romanelli took a Masters degree in design after graduating in architecture in 1983. He worked for Mario Bellini until 1985 when he became freelance. He was editor at *Domus* magazine from 1986 to 1994 and has been a design editor at Abitare since 1995. He works as a consultant with Driade and is art director at Montina and Oluce. Laudani and Romanelli currently work for Atlantide, Cleto Munari, Montina and Oluce. Their exhibition designs include the retrospective of Gio Ponti which was held at the 1997 Milan Furniture Fair. They have practices in Rome and Milan respectively and also work together on interior design projects. 132

Gaëlle Lauriot-Prévost is an architect and designer. He works in various fields from urban design to product design. Since graduation in 1991 he has worked extensively with the architect Dominique Perrault and in 1996 set up his own practice. Important works have included various projects for the French National Library. Recent work has comprised the competition for the City Hall of Marseille, the book *With* and lighting design for Fontana Arte – all in collaboration with Dominique Perrault. 122

Roberto Lazzeroni was born in 1950 in Pisa, Italy, where he still lives and works. After studying art and architecture in Florence, he began working in industrial and interior design. Since 1988 he has been art director of the Ceccotti Collections and he has collaborated with many companies, including Acerbis, Ciatta, Confaloneri, Driade, Moroso and Gervasoni. 122

Lemongras design studio was set up in Munich by Carmen Cheong and Moritz Engelbrecht. Both studied at the Royal College of Art, London, although they originate from Singapore and Germany respectively. They regularly exhibit in Milan and Cologne and have handled projects for Authentics and Deutsche Telecom. 140

Arik Levy was born in Tel Aviv and graduated from the Art Centre College of Design in Lausanne. He worked for a period in Japan before moving to Paris where he set up his own studio, 'L Design'. He took part in the 'Light Light' exhibition in Paris in 1998. 100, 109

Morten Linde was born in 1965 and studied at the School of Arts and Crafts, Copenhagen. For two years he was involved in the design of Bang & Olufsen products before opening his own architectural practice in 1992. He has worked for Tag Heuer, Kirk Telecom, Mexx Time, Grundig TV and Hewlett Packard, France. Most recently he designed a collection of 140 watches for Mexx Time which were presented at the Basle Fair 2000. 187

Frederick Lintz was born in France in 1971, graduating from ENSCI Les Ateliers in 1994. His preferred areas are product design, graphic design, illustration and architecture. After working for several design studios, he joined Design Plan Studio in 1997. 186

Piero Lissoni was born in 1956. He studied architecture at Milan Polytechnic and worked for G14 Studio, Molteni and Lema. He formed his own company with Nicoletta Canesi in 1984, involved in product, graphic, interior and industrial design and architectural projects. Since 1986 he has worked with Boffi Cucine as art director, creating corporate images and sales outlets, and in 1987 began to collaborate with Porro, Living Design, Matteograssi and Iren Uffici. He worked in Japan in the early 1990s for Takashimaya Company. Since 1994 he has worked as art director for Lema and in 1995 became art director for Cappellini, starting collaboration with Cassina and Nemo. In 1996 Lissoni was appointed art director for Units, the new Boffi and Cappellini kitchen company, and opened showrooms in Paris for Matteograssi and Boffi. He was awarded the Compasso d'Oro in 1991 for the Esprit kitchen designed for Boffi. In 1998 Lissoni moved to new premises and started collaborating with Benetton. Since then he has worked on various interior design projects including the headquarters of Welonda; two hairdressing salons; two new showrooms for Cappellini; the Allegri showroom; and the Boffi Bagni showroom in Milan. 40, 47, 60, 82

Mary Little graduated from the Royal College of Art, London. Her first collection of experimental upholstery 'Coat of Arms' was shown in London in 1994 and a piece from the collection was purchased by the Victoria and Albert Museum, London. Her work is in many private and public collections, including the Vitra Chair Museum and the Musée des Arts Décoratifs, Paris. In 1997 Mary Little and Peter Wheeler formed Bius. They design one-off limited editions and production furniture. 27

Ka-chi Lo was born in Hong Kong in 1962 and studied Electronics and Electrical Engineering at Middlesex Polytechnic, UK, followed by Industrial Design at Central Saint Martin's College of Art, London. After graduating in 1990 he mainly worked freelance for TKO Product Design and Graham Allen Associates. From 1998 he has worked as a freelance contractor for Papa Design and David Ames Design Studio and has won various awards. 129

Ross Lovegrove was born in 1958 in Wales. He graduated from Manchester Polytechnic in 1980 with a BA in industrial design, later receiving a Masters from the Royal College of Art, London. He has worked for various design consultancies including Allied International Designers, London and frogdesign in Germany. In 1984 he moved to Paris to work for Knoll International and became a member of the Atelier de Nîmes, a group of five designers which included Gérard Barrau, Jean Nouvel, Martine Bedin and Philippe Starck. In 1986 he co-founded Lovegrove and Brown Design Studio which was later replaced by Lovegrove Studio X. Clients include Louis Vuitton, Luceplan, Tag Heuer, Philips, Sony and Apple Computers. His work can be seen in major design collections, including MoMA, New York; the Guggenheim Museum, New York; the Axis Centre, Japan; the Centre Georges Pompidou in Paris and the Design Museum in London where, in 1993, he curated the first permanent collection. Lovegrove is a visiting lecturer at the Royal College of Art, London. Current projects include the Airbus A3XX, the first class advanced interior for Japan Airlines and a diversification programme for Tag Heuer in Switzerland. He is also designing a private villa for his family in Poland. 18, 30, 121

Henryk Lula is a sculptor, ceramicist and art restorer. Born in Poland in 1930, he studied at the Academy of Fine Arts, Gdansk, where he has held a chair since 1977. His works are held in the Polish national museums, the JFK Cultural Centre, Washington and ceramics museums in Faenza and Vallauris. He has won many awards and recent work has involved the restoration of the old city of Gdansk and Renaissance sculpture and ceramics in the Palazzo di Artus. 140

Massimo Lunardon, born in Marostica in 1964, is an artist and designer working primarily in glass. He received a Masters in industrial design from the Domus Academy and has collaborated with designers such as Andrea Anastasio, Ron Arad, Marc Newson and Javier Mariscal. He has exhibited throughout Europe, most recently at the Glass Biennale in Venice (1998/99) as well as at the New Glass Review at the Corning Museum of Glass in New York, 1997. He has also worked with Artemide, Flos, Driade and Memphis. 146, 147

Ane Lykke was born in Denmark in 1967 and graduated in industrial design from the Academy of Danish Design in 1996. She has worked for Design Studio Faro Designi in Rome, Bang & Olufsen and Jurgen Lehl Textile Studio, Tokyo. In 2000 she established a design studio in Copenhagen. 168

Kendo Makihara was born in Japan in 1955. He joined Sharp Corporation in 1974 as a product designer after graduating from the design school in Nagoya. He designs home appliances and in May 2000 was made head of GUI design. 198

Peter Maly is the head of Peter Maly Studio in Hamburg which works on product and interior design. The practice designs international furniture collections and also produces concepts and designs for trade fairs and exhibitions. Maly has received many awards and been the subject of various publications including a monograph by Form Verlag in 1995. Clients number COR, Behr, Interlubke, Thonet and Ligne Roset. He was born in 1936 and studied interior decoration at the Technical College, Detmold, after which he worked as a journalist and interior decorator for a German home service magazine. In 1970 he opened his design studio in Hamburg. 63, 72

Christophe Marchand was born in Friburgo in 1965 and studied industrial design in Zurich where he met Alfredo Häberli. From 1988 they have been curators at the Museum fur Gestaltung, Zurich and have worked with Alias since 1993. 81

Enzo Mari was born in Novara, Italy in 1932 and studied at the Academy of Fine Art in Milan. In 1963 he co-ordinated the Italian group Nuove Tendenze and in 1965 was responsible for the exhibition of optical, kinetic and programmed art at the Biennale in Zagreb. In 1972 he participated in 'Italy: the New Domestic Landscape' at MoMA, New York. Mari is occupied with town planning and teaching and has organized courses for the history of art department at the University of Parma and the architecture department at Milan Polytechnic. He has also lectured at various institutions including the Centre for Visual Communication in Parma and the Academy of Fine Arts in Carrara. He has been awarded the Compasso d'Oro on three occasions: for design research by an individual (1967); for the Delfina chair (1979) and for the Tonietta chair for Zanotta (1987). His work can be found in the collections of various contemporary art museums, including the Stedelijk Museum, Amsterdam; the Musée des Arts Décoratifs, Paris and the Kunstmuseum, Düsseldorf. Since 1993 he has collaborated with the KPM (Royal Porcelain Works) in Berlin, and in 1996 staged the 'Arbeiten in Berlin' exhibition at Charlottenburg Castle. 132, 148

Javier Mariscal was born in Valencia in 1950 and studied graphic design in Barcelona. In 1988 he created the mascot for the Barcelona Olympic Games and the following year founded Estudio Mariscal. Projects have included a new image for the Swedish Socialist Party (1993), covers for the *New Yorker* (1993/94/96) and the mascot for the Hanover 2000 exposition. He has designed furniture collections for Moroso and a bathroom collection for Cosmic as well as the corporate image for Barcelona Zoo. He also works in animation while his wide range of projects include furniture, textiles, porcelain and sculpture for such firms as Akaba, Adex, Memphis, Alessi, Swatch and Rosenthal. Mariscal also teaches throughout the world. 168

Mauro Marzollo was born in Venice in 1942. He studied industrial design at university and went on to work in Murano where he learned art techniques from famous master glass blowers. For some time he was head of the Design Department at Lumenform but has moved into private practice. He currently runs his own lighting design business, whose clients include Murano Due and Aureliano Toso 1938. 90

Jean-Marie Massaud was born in Toulouse, France, in 1966 and graduated from the Ateliers-Ecole Nationale Superieure de Creation Industrielle in 1990. He has worked for a range of European and Asian design consultancies. In 1994 he started his own studio concentrating on industrial and interior design for companies including Authentics, Baccarat, Lanvin, Magis and Yamaha. He has received several awards and his work is on show in the major museums of Amsterdam, Chicago, London, Paris and Zurich. 128

Leona Matejkova see Gabriela Nahlikova

Ippei Matsumoto was born in Japan in 1973 and studied industrial design at Tama Art University. In 1997 he started to work for IDEO Japan and has received awards from IF Hannover, IDSA and ID. 207

Ingo Maurer is widely acclaimed for his innovative and beautiful lighting designs. Born in 1932 on the island of Reichnau, Lake Constance, Germany, he trained in typography and graphic design. In 1960 he emigrated to the United States and worked as a freelance designer in San Francisco and New York before returning to Europe in 1963. He founded Design M in Munich in 1966 and since then has achieved worldwide recognition. He has exhibited widely and his works are in the permanent collections of major museums including MoMA, New York. He edited the 2000 edition of *The International Design Yearbook*. 96–7

Alberto Meda was born in Italy in 1945 and studied mechanical engineering at Milan Polytechnic. In 1973 he took up the position of technical director at Kartell. From 1979 he was consultant engineer and designer for Alias, Brevetti, CSI Colomo Design and Swatch Italia. He was a project director at Alfa Romeo from 1981 to 1985 and Professor of Production Technology at the Domus Academy, Milan form 1983 to 1987. Since he began working for Alias in 1987, three of his chair designs have selected for the collection of MoMA, New York and he has been awarded the Campasso d'Oro in 1989 and 1994. 33, 36, 68

Metalarte was founded in 1991 by Alberto Lievore, Jeanette Altherr and Manel Molina. Alberto Lievore was born in Buenos Aires, Argentina, in 1948 and studied architecture. After moving to Barcelona, he was involved in the set up of several influential design groups, including SIDI. Jeanette Altherr was born in Heidelberg in 1965 and in 1989, after graduating in Industrial Design, she began work with Alberto Lievore. Born in 1963, Manel Molina is from Barcelona and after graduating in Interior and Industrial Design he worked as a freelance designer. Metalarte is currently working for companies throughout Europe including Thonet, Disform, Perobell and Arper. They work in interior design, packaging design, product design, consultancy and art projects. 118

Vanessa Mitrani was born in 1973 and studied at the Ecole Superieure Internationale D'Administration des Enterprises and the Ecole Nationale Supérieure des Arts Décoratifs, specializing in furniture design. In 1999/2000 she spent a semester at the National Institute of Design, Ahmedabad, India, as part of an exchange programme. Professionally, she has been active in graphic design and in 1998/1999 worked with Pedro Veloso, the master glass-maker. Recently she has been working with Ligne Roset. 151

Luigi Molinis was born in Udine in 1940 and studied architecture in Venice. He started working for Zanussi in 1969 and was head of industrial design in he electronics department for ten years. Since 1980 he has worked freelance and is based in Pordenone with a client list which includes Zanussi, Radio Marelli, Ceramica Dolomite and Rhoss. 198

José Rafael Moneo was born in Navarra, Spain, in 1937 and graduated from Madrid's Technical School of Architecture in 1961. In 1970 he was made Professor of Architectural Theory at the Technical School of Architecture in Barcelona, where he taught until his return to Madrid in 1980. In 1984 he was appointed chairman of the Architecture Department at Harvard University and in 1990 named Josep Luis Sert Professor in Architecture. Among his best-known works are the Pilar and Joan Miro Foundation in Majorca, the Davis Art Museum, Massachusetts, the Museum of Modern Art and Architecture in Stockholm, the Auditorium in Barcelona and the Kursaal Auditorium and Congress Centre in San Sebastian (1999). 54

Mauro Mori was born in Cromona, Italy, in 1965, and lives in Parma. He trained in architecture but works mostly in carving, especially in wood. Inspiration is gained from frequent travelling and recent clients include Cappellini. 20, 38

Jasper Morrison was born in London in 1959. He studied design at Kingston Polytechnic, the Royal College of Art, London and the Hochschule der Kunste, Berlin. In 1986 he set up his Office for Design in London. He has designed furniture and products for companies including Alessi, Alias, Cappellini, Flos, FSB, Magis, SCP, Rosenthal and Vitra. In 1995 his office was awarded the contract to design the new Hannover tram for Expo 2000. The first vehicle was presented to the public in 1997 at the Hannover Industrial Fair and was awarded the IF Transportation Design Prize and the Ecology Award. Recent projects include furniture design for Tate Modern, London. Morrison's designs are in the collections of museums worldwide, including MoMA, New York. 39, 120, 171

Benny Mosimann was born in Baden in 1966 and studied interior and product design at the Zurich Design School. From 1990 to 1993 he was senior designer at Greutmann Bolzern. He then spent a year travelling around the world before studying graphic design at Basel Design School. In 1998 he founded his own studio whose focus is product design, graphic design and fair and exhibition design. 76

Pascal Mourgue was born in 1943 in Neuilly-sur-Seine, Paris and is a graduate of the Ecole Nationale Supérieure des Arts Décoratifs. He started his career at Prisunic. During the 1980s, when he was working independently, he started long-standing collaborations with several companies including Fermob, artelano, Roset, Toulemonde-Bochart, Scarabat and Cassina. He has since designed for Cartier and Baccarat; private homes for clients; shops for Roset in Chicago, Miami and Munich; and graphics for the Musée de la Poste in Paris. He has exhibited widely within Europe and in 1984 was elected 'Designer of the Year' by the Salon du Meuble de Paris. He received the Grand Prix de la Creation de la Ville de Paris in 1992 the Grand Prix de la Critique du Meuble Contemporain (1996) and the Grand Prix National de la Creation Industrielle (1996). Mourgue's work is in the permanent collection of the Musée des Arts Décoratifs in Paris. He produces sculptures, many of which have been exhibited worldwide. 26, 60, 168

Gabriela Nahlikova has worked with Leona Matejkova since 1996. Both attended the Academy of Art, Architecture and Design in Prague, studying in the design studio of Borek Sipek. Nahlikova spent a further year at the Academie Bellende Kunsten in Maastricht. They have exhibited widely. 27, 28, 74

Paola Navone was born in Turin in 1950. She graduated in architecture from Turin Polytechnic, her dissertation being published by Bruno Orlandini as *Architettura Radicale*. She worked for Alessandro Mendini, Ettore Sottsass and Andrea Branzi and as part of the the experimental groups 'Global Tools' and 'Alchimia' for whom she arranged major cultural events at the Venice Biennale of 1980. She was consultant art director to Centrokappa form 1975 to 1979 and has worked as a researcher at Centro Domus. She has acted as a consultant to Abet Laminati where she was responsible for market studies, product development and communication strategies. Other clients include Alessi, Knoll International and Fiat Auto. In 1985 she founded the manufacturing company Mondo with Giulio Cappellini. 20, 130

Marc Newson was born in Sydney, Australia, where he studied jewellery and sculpture. He started experimenting in furniture design at college where he staged his first exhibition. In 1987 he moved to Japan and in 1991 set up a studio in Paris in 1991, working for clients such as Cappellini and Moroso. He formed a joint venture Ikepod Watch Company which manufactured a range of watches as well as aluminium furniture such as the 'Event Horizon' table and 'Orgone' chair. Since the mid-1990s he has become increasingly involved in interior design and was responsible for restaurants such as Coast in London, Mash & Air in Manchester and Osman in Cologne. In 1997 he moved to London and set up Marc Newson Ltd; he has since become involved in wider range of mass-manufactured items for Alessi, Iittala and Magis. He has also designed the livery and interior of the $40 million Falcon 9008 long range private jet. His designs have won many prizes and can be seen in major design collections around the world. A monograph published by Booth-Clibborn Editions was published in 1999. 128, 160, 223

Nucleo is a team of young designers based in Torino, Italy. Now four years old, they concentrate on product and industrial design, with the aim of designing items which allow clients to use their own sense of creativity. 144

On Industriedesign is a German design team which consists of Klaus Nolting and Andreas Ostwald. Both were born in 1964 and studied at the Fachhochschule Kiel. They have worked for various design bodies. On Industriedesign's clients include Rolf Benz, Classicon, FSM Frank Sitzmobel, Hoffmeister Leuchten and Pieper Concept. 124

One Foot Taller are Katarina Barac and Will White. Both studied product design at Glasgow School of Art and are based in Scotland. They exhibited in New York in 2000. 30

Oz Design was founded in 1997 by Ely Rozenberg, Michael Garelick and Alessandro Bianchini who were born in Russia, Israel and Italy respectively. The group is dedicated to design which uses only the latest materials and technologies. They exhibited for the first time in Milan in 1998. Rozenberg studied industrial design at the Bezalel Academy of Art and Design, Jerusalem. He is the design correspondent on the *Biniyan ve Diur* (interior design magazine) and in 1996 was awarded first prize by 'Radad – the Israeli Cutlery Industry. Garelick also graduated in industrial design from Bezalel and then worked as head designer in the Bible Lands Museum, Jerusalem. He has been awarded the American Israel Foundation 'Sharet Grant' on three occasions and has also received the Brandeburg Prize for his innovative furniture designs. Bianchini studied architecture and industrial design in Venice and Rome and works on interior design projects. Since 1994 he has collaborated with various companies in Italy including Soft Line for whom he designed the 'Igloo' bed. 137

Satyendra Pakhalé was born in India in 1967 but has been based in Europe since 1992. He studied industrial design in Bombay and then at the Art Centre College of Design in Switzerland. From 1995 to 1998 he was senior product designer at New Business Creation, Philips Design. He worked on a concept car which was a collaboration between Philips and Renault. In 1998 he established Atelier Satyendra Pakhalé in Amsterdam. He is active in fields ranging from industrial design to furniture, crafts to interior architecture, and his client list includes Curvet, Habitat, Magis, Mexx International, and Zeritalia. 157

Ole Palsby was born in Copenhgen in 1935 and worked in finance until 1960 when he became a producer of home furnishings; in 1964 he opened an art gallery. Until 1968 he worked with a graphic designer, developing new marketing concepts and advertising programmes for design projects. In 1975 he began working as an independent designer and in 1986 opened a studio in London. He won both the Stuttgart Centre of Design Prize and the Japanese Design Prize. He has worked for many prestigious companies and won the If Product Ecology Design Award 2000. His work is held in museums and design centres worldwide. 136

Verner Panton was born in Denmark in 1926 and studied architecture at the Royal Art Academy in Copenhagen between 1947 and 1951. From 1950 to 1952 he worked in Arne Jacobsen's architectural studio, opening his own studio in 1955. The winner of several important awards for design, he created the Panton Chair for Vitra in 1969, the Swatch Art Clock Tower in Lausanne (1996) and an interior for Erco Lighting, London (1997). A fellow of the Royal Society of Arts, Panton died in 1998. Major retrospectives were held at the Design Museum in London in 1999 and at the Vitra Design Museum, Weil am Rhein in 2000. 37

Jiri Pelcl was born in 1950 in Czechoslovakia. He studied architecture at the Academy of Applied Art in Prague and furniture design at the Royal College of Art, London. He is the founder of the design Group ATIKA and in 1990 set up his own studio, Atelier Pelcl. Major commissions include Vaclav Havel's study in Prague Castle, St Laurence Church, Prague and the Czech Embassies in Rome and Pretoria. He has exhibited in galleries throughout the world and has been head of the School of Architecture and Design at the Academy of Applied Art in Prague since 1997. 80

Maurizio Peregalli was born in Varese in 1951. He studied in Milan and began to work on fashion shops and showrooms including the Giorgio Armani boutiques in Milan and London as well as on the image of the chain Emporio Armani. In 1984 he founded the furniture collection Zeus where he works today as Partner and Art Director. He is also a Director of Noto which produces Zeus. 111

Dominique Perrault is an architect, known for his refined and austere high-tech style. Born in Clermont-Ferrand, France, in 1953, he studied at the Ecole Supérieure des Beaux Artes, the Ecole Supérieure des Ponts et Education and the Ecole des Hautes Etudes en Sciences Sociales, graduating in 1980. Notable works include the Bibliothèque Nationale, Paris; the Ecole Supérieure, Marne-La-Valée; and the Olympic Velodrome, Berlin. 122

Christophe Pillet was born in 1959 and studied in Nice and the Domus Academy, Milan. After graduating, he collaborated with Martine Bedin and Philippe Starck but has been working alone since 1993. He was named Designer of the Year in 1994 and is active in furniture design, product design, interior design, architecture and set design. His clients include, Cappellini, Moet et Chandon, Lancôme, Bally and Moroso. 119, 212

Giancarlo Piretti was born in Bologna in 1940. He studied and later taught interior design at the Institute of Art in Bologna after which he spent 12 years as a furniture designer for Castelli SpA. In 1998 he launched the Piretti Collection, which is a highly successful office seating programme. He has twice won the Compasso D'Oro, and in 1997 launched the Torsion Collection. 36

Paolo Piva was born in Italy in 1950 and studied architecture in Venice. He also studied in Vienna, where he now teaches at the Academy Of Applied Arts. He has designed for renowned furniture companies and for buildings. Since 1974 he has had a close association with Wittmann GmbH, Germany. 44

Tim Power is an American architect/designer who has been living and working in Milan since 1990. Born in California in 1962, he studied architecture at the California Polytechnic State University. Having spent periods in the studios of Sottsass Associati, Superstudio and I.O.O.A., he opened his own studio in 1995. Clients include Zeritalia, Cassina/Interdecor, BRF, Poltronova, WMF, Rosenthal, and Fontana Arte. Recently his practice has begun expanding into interior and architecture work for clients in the art, media and advertising worlds. 66

Christopher Procter and Fernando Rihl both studied at the Architectural Association in London. They are well-known for their experimental work in the use of live-edge acrylic in furniture design. Before collaboration, Procter worked with Rick Mather and Paolo Solari and the engineer Tim McFarlane; and Rihl collaborated with the landscape architect Burle Marx. They worked with Zaha Hadid on panels for the Interbuild Blueprint Pavilion and in 1996 were commissioned to design acrylic partitions for five shops in central London. Procter and Rihl's work can be seen in the design collection of MoMA, New York. They are currently designing for industry with projects underway with North American, Swedish, Italian and Brazilian companies. They are also working on several residential and commercial projects in London and a beach house in Brazil. 79

Emaf Progetti was set up in 1982. Its chief activites are design, installations and communication for furniture makers. Its products have won awards and are owned by various museums. 48

Emma Quickenden graduated from Kingston University's Product and Furniture course in 1999. Since then she has exhibited in London, New York and Milan and is now working freelance. Her designs reinterpret aspects from traditional domestic living. 221

Karim Rashid was born in Cairo in 1960 and graduated in industrial design from Carleton University in Ottawa, Canada in 1982. After graduate studies in Italy he moved to Milan for a one-year scholarship in the studio of Rodolfo Bonetto. On his return to Canada he worked for seven years with KAN Industrial Designers in Toronto. Rashid was a full-time associate professor of industrial design at the University of Arts in Philadelphia for six years and has also taught at the Pratt Institute, the Rhode Island School of Design and Ontario College of Art. Since 1992 he has been principal designer for Karim Rashid Industrial Design in New York, designing products, lighting, tableware and furniture. He has won many awards including the 1999 George Nelson Award as well as the Silver IDEA Award for the Oh Chair. His work has been exhibited in museums internationally, including MoMA, New York; the Chicago Athenaeum; and the Design Museum, London. He recently held a show at the Sandra Gering Gallery called '5 Senses', exhibiting 120 limited edition ceramic sculptures and 3 Blobject sculptures. 18, 32, 98, 99, 152

Emmanuele Ricci was born in Treviso, Italy, in 1963. He studied at Milan Polytechnic and spent a period with the architect Gianfranco Frattini, after which he set up as an independent designer. He divides his time between a studio in Milan and the Treviso hills where he carries out artistic research. Among past and present clients are Alfa Romeo SpA, Artemide SpA, Fiat SpA,Chrysler, Lorenzo Rubelli SpA, Cassina, Dolomite, Lumina and Studio Thun. 166–7

Paolo Rizzatto was born in Milan in 1941 and graduated in architecture from Milan Polytechnic. He founded Luceplan in 1978 with Riccardo Sargatti, and from 1985 to 1987 he designed for Busnelli and Molteni and was also involved with interior architecture, planning and exhibitions, and interior design for private residences. Today he works as a freelance designer. He has collaborated with many leading manufacturers and has exhibited his work worldwide. Examples can be seen in the permanent collection of MoMA, New York. In 1990 he was invited to Japan to represent Italian design in the exhibition 'Creativitalia' in Tokyo. He has been awarded the Compasso d'Oro on three occasions: in 1981 for his lamp D7, in 1989 for the Lola lamp series produced for Luceplan and in 1995 for the Metropoli lamp series, again for Luceplan. Rizzatto has taught at various university institutes including Columbia University in New York, Milan Polytechnic, Washington University in Saint Louis and the Cranbrook Academy of Art in Michigan. 45, 68

Hannes Rohringer was born in Austria. He lives and works as a designer/artist in Seewalchen and Vienna, where he graduated with an MA in Fine Art from the Hochschule für Angewandte Kunstler. In 1989 he founded the Atrium studio which specializes in architecture, product design and applied art. His work is in the collections of various museums and his client list includes Duravit, Miele, Porsche Design, Molto Luce, Streitner and DCW Software. 77

Marco Romanelli see Marta Laudani

Heinz Röntgen worked as a freelance designer and consultant for textile companies during the 1950s. Since 1964 he has been marketing his work through his company Nya Nordiska. Recently he has focused on design and product development of interior fabrics. He has won many awards, including the Red Dot Award from Westfalen for the last decade, and the IF Product Design Award (1997, 1998, 2000). His work is held by the Chicago Athenaeum. 172, 173

Ely Rozenberg was born in the Soviet Union in 1967 and emigrated to Israel in 1977. Rozenberg studied industrial design at the Bezalel Academy of Art and Design, Jerusalem. He is the design correspondent on the *Biniyan ve Diur* (an interior design magazine) and in 1996 was awarded first prize by 'Radad' – the Israeli Cutlery Industry. 25, 136

Petra Runge is a graphic artist and designer who lives and works in Cologne. She was born in 1957 in Hannover and studied art and psychology. From 1987 to 1993 she worked mostly in France, on multidisciplinary cooperative projects. From 1994 she has worked as a graphic artist and designer for companies such as De Padova srl. 82

Joan Gaspar Ruiz is an industrial designer who lives and works in Barcelona. He was born in 1966 and studied at the Escuela de Artes y Oficios Artisticos in his native city. He developed his first lighting products for Vapor and since 1991 has worked for Disform, Santa & Cole, Sellex, Eurast and B-Lux. In 1996 he began working for Marset Illuminacion as the development and design director and also teaches at the Industrial Design School in Barcelona. 124

Junichi Saitou was born in Japan in 1965. He studied industrial design and joined Sharp Corporation in 1985, where he has designed video products. 183

Kasper Salto was born in Denmark in 1967 and qualified as a cabinetmaker in 1988. In 1994 he graduated from the Danish Design School. Since 1996 he has taught at the Royal Academy of Fine Art in Copenhagen while his furniture designs have won several awards including the *ID* prize in 1999. 33, 42

Thomas Sandell was born in 1959 in Finland. In 1981, after National Service, he studied architecture in Stockholm and in 1989 set up his own practice. He has worked on many prestigious interior and product design projects for clients such as Cappellini, IKEA, KLM, Swiss Air, Kallemo, B&B Italia, the Modern Museum in Stockholm and Ericsson. He has taught widely and as a winner of many awards is represented in both British and Swedish museums. 174

Santos and Adolfsdóttir was founded by Margaret Adolfsdóttir and Leo Santos-Shaw who met while studying at Middlesex Polytechnic, UK. Before forming their practice, Adolfsdóttir worked as a freelance fashion textile designer selling through agents in the USA to Calvin Klein and Perry Ellis. Santos-Shaw worked as a full-time assistant jeweller to Pete Change and then worked on a freelance basis selling to the USA, Europe and Japan. He was an in-house designer for Thierry Mugler in Paris. Santos and Adolfsdóttir produce interior textiles and surface design. Clients include Barneys, New York and Japan, and Whistles, London. They produce and market a range of scarfs which sell in Germany, Austria and at selected outlets in London. 174

Patrizia Scarzella is an architect and journalist. She was on the editorial staff of *Domus* from 1980 to 1986 and worked with leading design firms. The author of several books, she was Zanotta's corporate image and communications consultant from 1985 to 1996. 56

Jürgen Schmidt was born in 1956. He established Design Tech in 1983 and is the author of several books. He has won the Braun Award for technical design and the VDID award of design for the disabled. He is a member of the iF Hannover 2003 design jury. 205

Henrik Schulz is a member of the Copenhagen-based group -ing. He studied architecture in Gothenburg, design in Copenhagen and architecture at the Royal Danish Academy of Fine Arts, graduating in 2000. Schulz has exhibited in Stockholm, Copenhagen, London and Milan. 24

Iwan Seiko was born in 1951 in Croatia and studied philosophy in Zagreb. He now lives in Munich and works as a painter, artist and sculptor. 94

Kazuyo Sejima is an internationally-known architect. She was born in Japan in 1956 and graduated in 1981 in architecture from the Japan Women's University. She joined Toyo Ito and Associates but set up her own practice, Kazuyo Sejima and Associates, in 1987. Her work is characterized by its poetic tone combined with modern architecture and the use of materials like glass, aluminium, metallic grilles and plastics. She has won many international awards and famous works include the Platform House series, the N-House, the Y-House and the Villa in the Forest. Recent projects include the O-Museum, Usika New Station Building, the Museum of Contemporary Art of Sydney Extension and the restoration of the historic centre of Salerno, Italy, for which she won a prize in 1999. 51

Richard Shemtov graduated from Parsons School of Design in 1990. A native of New York, he founded the International Design Supply Corporation (IDS) in 1991 to showcase his work in interior and furniture design. His interior design work has included a spa, a restaurant and the Madison Avenue branch of Dolce & Gabbana. In 1996 he founded Dune, designing furniture and prototypes for retail and individual designers and in 1998 launched his own furniture line at the ICFF. In 1999 he formed a collaboration with other American designers, Nick Dine, Jeffrey Bernett and Harry Allen and is expanding Dune. 61

Asahara Sigeaki was born in Tokyo in 1948 and studied in Torino, Italy. Since 1973 he has worked as a freelance industrial designer and in 1992 received the 'I.F. Best of Category' prize at the Hannover Exhibition. He is active between Italy and Japan and is permanently represented in the Brooklyn Museum of New York. 121

Tuttu Sillanpää was born in Finland in 1967 and graduated with an MA from the University of Art and Design, Helsinki. In addition to teaching art and textile design at Helsinki Secondary School of Visual Arts, she has worked as a designer for the Italian company Ratti SpA, and since 1998 the Finnish company Verso Design. Her work has been exhibited in Italy, France, the UK and Scandinavia. 174

Claudio Silvestrin was born in 1954 and trained in Milan at AG Fronzoni. He completed his studies at the Architectural Association in London where he now lives and works. He teaches at the Bartlett School of Architecture in London and at the Ecole Supérieure d'Art Visuels in Lausanne. Some of his most important works include shops for Giorgio Armani (Paris), the offices, shops and home for Calvin Klein (Paris, Milan and New York) and works for museums and art galleries. 40, 214, 215

Borek Sipek was born in Prague in 1948. He studied architecture in Hamburg, philosophy in Stuttgart and architecture in Delft. In 1983 he moved to Amsterdam and started his own studio. Projects include the Tsjeck Pavilion for the World Exhibition 2000 in Hannover, objects for Driade, Scarabas and Vitra. He received La Croix Chevalier dans l'Ordre des Arts et des Lettres from the French government in 1992 and in the same year was appointed Court Architect in Prague. He is currently engaged on projects in Prague. 130, 142, 149

Barbora Skorpilova was born in 1972 and studied architecture in Prague, under Borek Sipek. Since graduating in 1997 she has worked with Jiri Pelcl. From 1994 she has worked with *Elle Décor*, *Esquire* and *Harper's Bazaar* and, since 1996, with Jan Nedved (b.1973) who also studied under Borek Sipek. In 1999 they founded Studio Mimolimit together. 140

Smart Design's Director of Design Engineering is Clay Burns. He graduated from Dartmouth College in 1987, and took an MA at Tufts University in 1989. He began his career with Product Genesis, Massachusetts and then moved to New York where he worked as an independent consultant for some time, specializing in product engineering and ergonomics. Recent projects include development of technologies for athletic footwear, cushioning, research and testing of new household firesafes and improvement of grip and vibration attenuation for hammer handles. 210

Snowcrash design team is based in Stockholm and has emerged as one of the most innovative design companies of recent years. It has explored new material and technologies to great effect in conjunction with the information and communications boom and regularly exhibits in Milan. 70, 100

Michael Sodeau studied product design at Central Saint Martin's College of Art and Design. He was a founder partner of Inflate in 1995 but left in 1997 to set up a new partnership with Lisa Giuliani. He launched his first collection of furniture and homeware, Comfortable Living, in 1997. 169

Michael Solis was born in Dallas, Texas, but now lives and works in New York where he runs his own furniture/product studio, WORX. He graduated from Parsons School of Design in 1991. In 1995 he launched his first line of furniture and home accessories at the New York International Contemporary Furniture Fair and in 1996 won the Editor's Award for Best New Designer. Recently he has been working with Nick Dine and also shows at Totem, New York. His work has been included in publications worldwide. 61

Peter Solomon graduated from Pratt Institute, New York in 1988. Until 1993 he lived in California, designing museums, computers, furniture, lighting, toys and interiors. He then moved to Milan to obtain an MA at the Domus Academy and subsequently established a studio. For over six years he has collaborated with Isao Hosoe and designs a range of products from telephones to furniture, lighting and sporting equipment to interiors. 93

Ettore Sottsass was born in Innsbruck, Austria, in 1917, and graduated in architecture from Turin University in 1939. In 1947 he opened an architecture and design studio in Milan. In 1958 he began working with Olivetti as a consultant, designing the first Italian computer in 1959. In 1981 he and various colleagues formed Memphis. The following year he founded Sottsass Associati, where he still works as an architect and designer. Sottsass holds honorary degrees from the Royal College of Art, London and the Rhode Island School of Design while the Centre George Pompidou, Paris held a major retrospective of his life and works in 1994. His works form part of the collection of major museums worldwide and he is internationally recognized as a giant of innovative design. 79, 154

Hanspeter Steiger was born in Switzerland in 1970. He served an apprenticeship as a furniture-maker before undertaking study at the Schule fur Gestaltung, Basel, the National College of Art and Design, Oslo and the Danish Design School in Copenhagen. He has exhibited in Milan and Copenhagen and won several prizes for his work. Steiger is also a member if the -ing group. 24

Philippe Starck was born in Paris in 1949 and trained at the Ecole Camondo in Paris. After a spell in New York he returned to France where he has built up an international reputation. He has been responsible for interior design schemes for François Mitterrand's apartment at the Elysée Palace, the Café Costes, and the Royalton and Paramount Hotels in New York. He has also created domestic and public multi-purpose buildings such as the headquarters of Asahi Beer in Tokyo, the Ecole Nationale Supérieure des Arts Décoratifs in Paris, and the air traffic control tower for Bordeaux Airport. As a product designer he collaborates with Alessi, Baleri, Baum, Disform, Driade, Flos, Kartell, Rapsel, Up & Up, Vitra and Vuitton. From 1993 to 1996 he was worldwide artistic director for the Thomson Consumer Electronics Group. Awards include the Grand Prix National de la Création Industrielle and his work can be seen in the permanent collections of all the major design museums. In 1997 he completed hotels in Miami and Los Angeles, and in 1998 the Canary Riverside Hotel in London, a hotel in Bali, the restaurant in the Hilton Hotel in Singapore and an incineration plant in Paris/Vitry. In 1999/2000 he finished two central London hotels for the Ian Schrager group and is now working on hotels in New York, San Francisco and Santa Barbara. 32, 101

Norma Starszakowna is Director of Research at the London College of Fashion. She has taught widely and sat on many advisory boards, including those of the Design Council and the British Academy. She has been closely involved in the production of textiles for public and commercial bodies including Issey Miyake; Nuno Co, Tokyo; Fitch and Co; General Accident Assurance Co; and the Scottish Arts Council. Her work has been exhibited widely in the UK and abroad, is held in many private collections and she has been the recipient of various awards. 165

Asher Stern was born in Jerusalem in 1972. After completing a diploma in jewellery making he studied at Bezalel Academy of Art and Design, Jerusalem, graduating in 2000. In the same year he exhibited in Milan. 205

Stile Bertone SpA was founded in 1971 by the Nuccio Bertone. Bertone itself was founded in 1912. In 2000 Stile Berone and Tecno design formed the Bertone Design Machine, a team of specialists utilizing state of the art technologies. They are involved in vehicle design and industrial design, from fork lift trucks to dental chairs, aircraft interiors to mountain bikes. 223

Reiko Sudo was born in Ibaraki Prefecture, Japan, and educated at the Musashino University of Art. From 1975 to 1977 she assisted Professor Tanaka in the textile department. Before co-founding Nuno Corporation in 1984, she worked as a freelance textile designer and has since designed for the International Wool Secretariat, Paris and Threads, Tokyo. She is the director of Nuno Corporation and a lecturer at the Musashino University of Art. Her work can be seen in the permanent collections of MoMA, New York; the Cooper-Hewitt National Design Museum, New York; the Museum of Art, Rhode Island School of Design; the Philadelphia Museum of Art; the Museum of Applied Arts, Helsinki; and the Musée des Arts Décoratifs, Montreal. Recent exhibitions include the 'Tokyo Creation Festival', Tokyo, 'The Textile Magician' show at the Israel Museum of Modern Art; 'Japanese Textile Design' at the Indira National Centre for Arts in India and 'Structure and Surface: Contemporary Japanese Textiles' at MoMA, New York. She has received many prizes for her work, including the Roscoe Award in 1993 and 1994. 176–7

Shinichi Sumikawa was born in Tokyo in 1962. After graduating from Cdhiba University's Industrial Design Department in 1984, he joined Sony as a product designer working on Walkmans, radios, headphones and TV. In 1992 he established Sumikawa Design in Tokyo and has been designing communication tools, cars, medical equipment and sports gear. 206

Marco Susani and Mario Trimarchi are architects and industrial designers and have been working together since 1986, developing various projects including house appliances, kitchen tools, lamps, furniture and corporate identity. They are responsible for the art direction fro Serafino Zani, also developing its new corporate identity. 143

Martin Szekely is one of France's best-known designers. He has exhibited in Europe, the US and Japan and clients include: Swarovski, Cassina, Hermès, Perrier and Montina. He is working for Hermès, SNCF and Dom Pérignon. Szekely's work is in the collections of major museums in the US, Germany, France and Israel; in 1998 he was included in 'Premises', an exhibition of the four most famous French designers at the Guggenheim Museum, New York. He was awarded the title of Chevalier of Arts and Letters in 1999. 83, 220

Tomoki Taira was born in Japan in 1971. He studied industrial design in Osaka, graduating in 1991. He then joined Sharp and is in charge of product design development at the Corporate Design Centre and Appliances Design Centre. 199

Yumiko Takeshita was born in Japan in 1962. In 1985 she graduated in Industrial Design from the Women's University of Art and joined Sharp Corporation. Her preferred field is home appliances and since 1995 she has worked at the corporate design centre in charge of advance design and special projects. 183

Carlo Tamborini was born in Milan in 1958 and studied there at the European Institute of Design. After graduation in 1993 he worked freelance on interior design projects. From 1997 to 2000 he worked with Lissoni Associati and in 1998 became a member of Codice 31, concentrating on product design. Recent projects have included product design for Palluco Italia and lighting for Fontane Arte. 82

Pepe Tanzi was born in Monza, Italy, in 1945. He graduated from Milan Polytechnic and today works for Album as industrial designer on their range of lighting products. He also works on a freelance basis for numerous leading lighting manufacturers. 69

Pascal Tarabay was born in Beirut in 1970. In 1998 he did an MA in design at the Domus Academy, Milan, under Andrea Branzi. The following year he worked as Andrea Branzi's project assistant and also for Donegani & Lauda and Aldo Cibic. Since 1993 he has been a freelance designer, numbering Salvatore Ferragamo, Radice, Driade and Beirut Municipality among his clients. He has won several prizes for his work and has exhibited in Milan and Cologne. 32, 42, 44

Matteo Thun was born in Germany in 1952 and studied in Salzburg, then Florence where he received his PhD. In 1981, together with Ettore Sottsass, he founded Sottsass Associati and the Memphis design group. From 1983 to 1986 he was professor of design at the University of Applied Arts in Vienna and since 1984 has lived and worked in Milan. 98

Jacob Timpe was born in Wurzburg, Germany in 1967. He studied architecture in Berlin and Darmstadt and worked as a freelance architect for various studios, winning several prizes. In 1999 he designed 'Tishbocktisch' for Nils Holger Moorman and is currently working as an assistant in the architectural department of the Technical University of Darmstadt. 60

Pekka Tiovola was born in Finland in 1955 and studied at the University of Industrial Arts in Helsinki. Since 1986 he has worked for Martela and is a leading figure in office furniture design. He has won several awards, including the Gold Clip at the International Fair in Prague, 1997. 66

Stephan Titz was born in Germany in 1971 and studied mechanical engineering before training as a cabinetmaker. From 1995 to 1997 he worked for MANUFORM Barbin and in 1998 completed an MA in Furniture design at the Savannah College of Art and Design. He currently works as a freelance designer for Natural Living of Austria and Team 7, while his work has been exhibited in New York, Cologne and Paris. 78

Frank Tjepkema was born in Geneva in 1970. He studied industrial design at Delft University and went to the Design Academy, Eindhoven where his course included some study at the Royal College of Art, London. In

1998 he completed an MA at the Sandberg Institute, Amsterdam. From 1996 to 1998 he worked as an independent designer for Olliiy and Pirelli among others. From 1998 to 1999 he worked as a junior art director for Bruggenwurth Maas en Boswinkel in Amsterdam whose clients include Adidas and Heineken. Since 1999 he has been working independently for several advertising, marketing and communication agencies in Amsterdam, the Switzerland-based Swatch group and Droog Design. 85

TKO Product Design was founded by Andy Davey in 1990. He graduated from the Royal College of Art and has since designed for NEC, Canon and Sony among others, as well as creating eyewear for Seiko, lights for Daiko and toys for Hasbro. Davey was one of the first British designers to establish successful business links with Japanese manufacturers. TKO also works for major companies in the USA and Europe. Davey and his team have won many design prizes including the 1996 BBC Design Award for Best Product, the BBC Designer of the Year and Best of Category in the consumer products section of *ID*'s Annual design review for the Freeplay clockwork radio. 186, 211

Kazuhiko Tomita was born in Nagasaki, Japan, in 1965. He gained a B.Eng. in industrial design at Chiba University and in 1990 won a Cassina scholarship to study furniture design at the Royal College of Art, London. He was awarded first prize in the 'Architectural Future of Stainless Steel' competition and the MA RCA Marchette Award for his degree work 'Hadaka-no Piano, aria'. He has exhibited frequently at the Milan Furniture Fair as well as Abitare il Tempo. 200

Luigi Trenti was born in Florence in 1965 and graduated from the Architecture University of Florence with a published and SMAU-awarded thesis on industrial design. He has worked as a product designer since 1989, specializing in lighting and instrumentation design. Clients include Targetti Sankey, Osram, General Electric, Martini, Radim Group and Pineider. In 1993 he was appointed chief of Industrial Design at Targetti Sankey who won the Compasso d'Oro in 1998. In 2000 he won the Young & Design Award. 208

Oscar Tusquets Blanca is an architect, painter and designer. His works are in various permanent collections around the world. He has received numerous prizes and awards for architecture and design projects including the Chevalier de l'Ordre des Arts et des Lettres. In 1994 he published his first book as an essayist and has since written a second book. 154

Shigeru Uchida was born in Japan in 1943 and graduated from the Kuwasawa Design School in Tokyo. In 1970 he established the Uchida Design Studio and since then has lectured worldwide. Commercial work has included a Seibu department store, the Wave Building, Roppongi, a shelf component system for Esprit, boutiques for Yohji Yamomoto and a clock for Alessi. Recent projects include the Kobe Fashion Museum, a hotel and an eyewear shop in Japan as well as design for Abet Laminati. He has exhibited widely and his work is held in many major design collections, including MoMA, New York. 55, 79

Paolo Ulian (b. 1961) studied painting in Carrara followed by industrial design in Florence. From 1990 to 1992 he worked as an assistant at Enzo Mari's studio, then founded Paolo Ulian Industrial Design. He has exhibited widely and collaborates with international companies such as Driade, Segno, Progetti and Zani & Zani. 60, 108, 204

Patricia Urquiola was born in Oviedo, Spain, in 1961. She studied architecture in Madrid and then at Milan Polytechnic, graduating in 1989. From 1990 to 1996 she worked with Vico Magistretti as a consultant for De Padova and from 1993 to 1996 she developed franchise projects with the architects Marta de Renzio and Emanuela Ramerino. In 1998 she became the design manager for Lissoni Associates, working with Cappellini, Porro and Cassina. Recent projects have included lighting, furniture and product designs for Moroso, Fasem, Bosa and Tronconi. 46

Jaap van Aarkel was born in The Netherlands in 1967. He studied at the Industrial Design Academy, Eindhoven and since graduating in 1996 has exhibited in Milan, Rotterdam and Paris with Droog Design. 104

Dick van Hoff was born in Amsterdam in 1971. He trained as a building worker and window dresser before studying three-dimensional design at the Hogeschool voor de Kunsten in Arnhem. In 1995 his tap Stop Kraan was taken up by Droog Design and he has since had several other pieces developed by them and DMD as well as Rosenthal and United Colors of Benetton. 221

Christine van der Hurd studied textiles at Winchester College of Art, UK. After graduation she designed fabrics and home furnishings for clients such as Kenzo, Mary Quant, Biba, Cacharel, Liberty, Osborne & Little and Courtaulds. In 1976 she moved to New York and in 1980 began to specialize in carpeting. In 1981 she established the contemporary furniture store 'Modern Age' with her husband and in 1990 set up Christine van der Hurd Inc. She returned to the UK in 1997 and in 1998 Modern Age was relaunched as 'Cappellini-Modern Age', a flagship store for the Italian company. As well as being president of this venture, van der Hurd designed an exclusive range of rugs for Cappellini, launched at the 1999 Milan Furniture Fair. Christine van der Hurd Inc. is now based in London. 176

Peter van der Jagt was born in The Netherlands in 1971 and studied 3D design in Arnhem. In 1996 he set up his own studio and has completed several projects for Droog Design and Authentics. His work has been displayed at exhibitions in Rotterdam, San Francisco, Berlin, Milan, Frankfurt, Cologne and London. In 1999 he won the Rotterdamse Designprijs and he currently works as a freelance industrial designer and creative consultant for several companies. 85

Marijn van der Poll was born in The Netherlands in 1973. After moving around the Middle and Far East his family moved back to The Netherlands in 1993. He studied fine arts and interior design/architecture and is currently studying under Gijs Bakker at the Design Academy, Eindhoven. 85

Jan van Lierde was born in The Netherlands in 1954. He studied architecture in Ghent and in 1978 set up his own practice, undertaking domestic and commercial projects in Belgium, Algeria, Egypt and Scotland. In 1983 he established Kreon in Beveren-waas, a company which offered advice and distributive trade in light fixtures and interior design. In 1984 Kreon moved to Antwerp and in 1986 exhibited the 'Metis' collection in Milan. 113, 133

Maarten Van Severen is an interior and furniture designer. He is mainly active in the design of small-scale domestic and retail schemes and has recently completed the Maison à Floirac with the architect Rem Koolhaas and the interior entrance hall of the City Hall in Ghent. His furniture and lighting designs have been in production since 1997 and he now works for clients such as Vitra, Switzerland, U-line Lighting and Target Lighting in Belgium. He has exhibited his work in group shows throughout Europe. In 1998 he received the IF Design Award, Hanover for U-line and the Flemish government's design prize. He is a visiting professor at the Academy of Fine Arts in Maastricht. 44, 80

Luc Vincent was born in 1952. After graduating in both Interior Architecture and Journalism, he worked on product development for companies such as Habitat, Casa and La Maison. From 1987 to 1992 he helped develop Barcelona Airport, the Swift Building, J.C Decaux Building, and the Rochas building as well as products for Cassina, Alessi and Cartier. For the past ten years he has owned an interior architecture bureau in Brussels which participated in the building of the European Parliament and has developed products for Modular Lighting Instruments, Obumex and Totem. 114–5

Roderick Vos was born in 1965 in The Netherlands and studied industrial design in Eindhoven. He worked for Kenji Ekuans GK in Tokyo and then for Ingo Maurer in Munich before founding his own studio, Studio Masaupertuus, with Claire Vos-Teeuwen. The client portfolio includes Espaces et Lignes, Driade, Authentics and Alessi while work has been shown at the Milan, Cologne and New York furniture fairs. 16, 17, 220

Norbert Wangen was born in 1962 in Prum, Germany, and served a carpenter's apprenticeship before studying sculpture and architecture in Dusseldorf, Aachen and Munich. In 1991 he graduated in architecture from the Technical University in Munich and worked as a set designer. In 1995 the folding armchair Atilla became part of the 'Die Neue Sammlung Munchen' and in 1997 became part of the Vitras Design Museum collection. The kitchen was presented at the Milan Furniture Fair in 2000. 216

Carol Westfall graduated from the Rhode Island School of Design in 1960 and Maryland Institute College of Art in 1972 after which she undertook diverse postgraduate study in the USA, Mexico, India and Japan. Currently a Professor at Montclair State University, New Jersey, her work has been exhibited worldwide, awarded various prizes and is held in many collections. 175

Hannes Wettstein works in product design, corporate design, interior design and architecture. He was born in Switzerland in 1958 and started off doing freelance work. He then joined the Eclat Design Agency as a partner. In 1993 he co-founded 9D Design and he also teaches, having held a professorial chair at the Academy of Design in Karlshule since 1994. He has won many awards and notable projects include the Swiss Embassy in Tehran and the Grand Hyatt Hotel in Berlin. His varied list of clients has recently numbered Cassina, UMS Pastoe, Shimano, Baleri Italia and Artemide. 26, 45, 72, 187

Peter Wheeler is a partner in the design studio Bius. His background is in designing furniture for manufacture while Mary Little designs one-offs. Bius design unique furniture to commission for public art, corporate and private clients. Its work is in many public collections, including Manchester City Art Galleries and Museums, UK and Museo de las artes decoratives, Barcelona, Spain. 27

Sachio Yamamoto was born in Japan in 1950 and graduated in industrial design from Musashino Art University in Tokyo in 1975. He then joined Sharp Corporation where he has been in charge of product design development in the field of home appliances, audio and communication products. Since 1999 he has been a chief designer in charge of advance design development. 183

Kazuhiro Yamanaka was born in 1971 and studied furniture design in Tokyo, then interior design at the Royal College of Art, London. In 1999 he won the 100% Design/Crafts Council Bursary Award. 112

Paolo Zani was born in Italy in 1960. He studied Industrial Design in Faenza and then took an MA at the Domus Academy, Milan, finishing in 1987. As an independent designer he has worked in many areas for many companies such as Ross El, Winterling and Bareuther, Optly Holding, Schopenhauer, Fontanaarte, Moroso and Mont Blanc. Since 1991 he has been a member of the Domus Design Agency and exhibits frequently in Europe and Japan. 155

Acquisitions by Design Collections in 2000

Dates given in parentheses refer to the dates of the designs (from 1960 to the present day).

AUSTRALIA

POWERHOUSE MUSEUM, SYDNEY, NEW SOUTH WALES

FURNITURE
Arthur Robertson, wall panels (1998)

TABLEWARE
Benjamin Edols and Kathy Elliot, bowl (1998)

TEXTILES
Jimmy Pike, rug, Jilji and Kurrminti (1987) manufactured by Prom Thai

PRODUCTS
Wayne Stuart, grand piano with stool (1998–9) manufactured by Stuart & Sons

AUSTRIA

ÖSTERREICHISCHES MUSEUM FÜR ANGEWANDTE KUNST, VIENNA

FURNITURE
Frank Gehry, chair, Grandpa Chair (1987)
Rupert Walser, picture/table (1982–5)
Manfred Erjautz, chair, Electric Chair (1996)
Liam Gillick, screen, Layered Impasse Screen (1998)

TABLEWARE
Achille Castiglioni, fruit bowl (1995)
Achille and Pier Giacomo Castiglioni, cutlery set (1997)
Michael Graves, cutlery set (1995)

CANADA

MUSÉE DES ARTS DÉCORATIFS DE MONTRÉAL

FURNITURE
Ali Tayar, table, NEA Table 2 (1994)
Masanori Umeda, cabinet, Ginza Robot (1982)
Robert Venturi, chair
Marcel Wanders, chair, Geknoopte Stoel (1998)
Afra Bianchin Scarpa, footstool, Siroana (1970–99)
Jonathan de Pas, sofa, Blow, (1967)
Tom Dixon, chair, Pylon (1991)
Frank Gehry, desk
Shigeru Uchida, shelves, Stormy Weather (1999)
Umeda Masanori, cabinet, Ginza Robot (1982)
Tejo Remy, chiffonnier, 'You can't lay down your memories' (1991)
Ettore Sottsass, sideboard, Beverly (1981)
Philippe Starck, stool, Bubu 1er
Marc Newson, chair, Wicker Chair (1990)
Andrea Anastasio, chest of drawers, Alba (1999)
Fernando Campana, table, Inflatable Table (1995)
Achille Castaglione, desk, Servolibro (1985)
Don Chadwick, office chair, Aeron (1992)
Don Chadwick, chair, Chadwick Modular Seating (1974)
Joe Colombo, wheeled cabinet, Boby (1970)

LIGHTING
Tejo Remy, chandelier, Milkbottle lamp (1997)
Ettore Sottsass, table lamp, Tahiti (1981)
Vico Magistretti, table lamp, Atollo (1977–99)
Ingo Maurer, table lamp, Samurai (1998)

TABLEWARE
Edmund de Waal, vases (1999)
Robert Turner, container, Ashanti (1999)
Mary Rogers, bowl (1975–9)
Toby Russell, jug
Kate Malone, jug (1999)
Sven Palmqvist, wine glasses
Andrea Anastasio, vase, Presente (1999)
Andrea Anastasio, vase, Ciliego (1999)
Constantin Boym, vases, American Plumbing

PRODUCTS
Giancarlo Piretti, umbrella, Pluvium (1970)
Akira Onuzuka, watches, Chewing Gum Series (1997)
Akira Onuzuka, watches, Le Chocolat Series (1997)
Akira Onuzuka, watches, Zoo Series (1997)

FRANCE

MUSÉE UNION CENTRALE DES ARTS DÉCORATIFS, PARIS

FURNITURE
Riccardo Blumer, chair, Lalaggera (1999), manufactured by Alias
Alberto Meda, chair, Highframe (1999), manufactured by Alias
Jean-Marie Massaud, chair, O'Azar (1999), manufactured by Magis srl
Marc Berthier, desk, Ozoo 700 (1997)

TABLEWARE
Francois Bauchet, coffee service, Vallauris (1989) manufactured by Edition Arcodif
Claude Bouchard, dinner service, Tara (1998), manufactured by Edition Arcodif
Pablo Reinoso, table d'appoint, Pocketable, manufactured by Edition Arcodif
Eric Jourdan, Pichet, Marguerite (1999) manufactured by Edition Arcodif
Olivier Gagnere, dessert cutlery, Ithaque (1985) manufactured by Bernardaude
Helen Von Boch, 19 piece service, Boule (1991) manufactured by Villeroy and Boch
Martin Szekely, dessert ware, Maria (1990) manufactured by Faiencerie de Gien

PRODUCTS
Tom Dixon, zinc-wire wall clock (1993)

GERMANY

KUNSTMUSEUM DÜSSELDORF IM EHRENHOF

PRODUCTS
Hi-fi, Concept 51k (1970) manufactured by Wega Radio GmbH
Electric razor, Rolan Ullermann, (1977) manufactured by Braun
Ettore Sottsass, calculator, GH. Von KLIER (1970) manufactured by Olivetti
Dieter Rams, Alarm Clock manufactured by Braun AG
Alarm Clock manufactured by Braun AG

VITRA DESIGN MUSEUM, WEIL AM RHEIN

FURNITURE
Pietro Arosio, chair, Mirandolina Chair, manufactured by Zanotta
Werner Aisslinger, chair, Juli Armchair, manufactured by Cappellini SpA
Ron Arad, chair, FPE Chair, manufactured by Kartell
Philippe Starck, chair, La Marie Chair, manufactured by Kartell
Philippe Starck, chair, Dr No Chair, manufactured by Kartell
Vico Magistretti, chair, Maiu, manufactured by Kartell
Antonio Citterio, chair, Dolly Chair, manufactured by Kartell
Roy Fleetwood, sofa, Wing, manufactured by Vitra
Verner Panton, chair, Tatami, manufactured by Kohseki Co.
Maarten van Severen, armchair and lounge chair
Ettore Sottsass, table, Mimosa, manufactured by Memphis
Ettore Sottsass, table, Cream, manufactured by Memphis
Carolyne Schlyter, chair, Lilla-h, manufactured by Swede
Shiro Kuramata, Sofa Table
Mario Bellini, dining table, Il Colonnato
Andrea Branzi, domestic animals couch, manufactured by Zabro
Ruth Francken, chair, Homme
Yonel Lebovici, chair and footrest, Clip
Bruno Rey, chair, Chair 33, manufactured by Dietker & Co
Verner Panton, glass chair and table, manufactured by Hebsgaard

LIGHTING
Verner Panton, lamp, Panto-Lamp, manufactured by Louis Poulsen
Yonel Lebovici, plug floor lamp, Male

PRODUCTS
Philippe Starck, toy, Big-the Face

THE NETHERLANDS

MUSEUM BOIJMANS VAN BEUNINGEN, ROTTERDAM

FURNITURE
Maarten Van Severen, trestle table (1998), manufactured by BULO

TABLEWARE
Jurgen Bey, tea service (1999)
Jan van der Vaart, drinking glasses (1995–9), manufactured by Glasfabrick Ajeto

PRODUCTS
Michele Grion, foldable coathanger (1999)
Ton Haas, Office Accessories (1998–9), manufactured by Fair Trade
Peter Van der Jagt, floor and wall tiles (1997), manufactured by Spinx
De Denktank, aluminium rulers (1995–7), manufactured by The Edge
Morison S. Cousins, Peeler (1997), manufactured by Tupperware Corporation
René Floore, Baby Carriage (1995), manufactured by OutDoorBabyCars

STEDELIJK MUSEUM, AMSTERDAM

FURNITURE
Richard Hutten, child's chair, Bronto (1999)
A+E Design, folding chair, Stockholm II (1996)
Marc de Vree, stool M1+ (1998)
El Ultimo Grito, chair, Miss Ramirez (1997)
Philippe Starck, chair, Dr No (1996)
Philippe Starck, chair, Miss Trip (1996)
Philippe Starck, chair, Super Glob (1991)
Antonio Citterio, Oliver Low, serving trolley, Batista
Ron Arad, bookshelf, Bookworm (1994)
Bjorn Dahlstrom, chair, BD-6 (1996)
Creadesign, foldable chair, Baby Trice (1985)
Jouko Jarvisalo, chair, Kova B (1997)
Stefan Lindfors, chair (1996)
Peter Karpf, chair, Eco (1997)
Valdimar Hardarsson, chair, Soley (1984)
Peter Opsvik, chair, Conventio (1997)
Peter Opsvik, chair, Gravity (1983)
Niels van Eijk, chair, Cowchair (1997)

LIGHTING
Mario Bertorelle, lamp (1970s)

TABLEWARE
Sigurd Bronger, pepper and salt set (1996–7)
Erik Magnussen, service, Stelton (1977)
Ingegerd Raman, service, Strof (1997)
Ingegerd Raman, service, Gronstedt (1997)
Ingegerd Raman, service, Hesselbom (1997)
Ingegerd Raman, decanters and glasses (1997)
Stefan Lindfors, service, Ego (1997)
Stefan Lindfors, bowls from the Tools series (1998)
Kati Tuominen-Niityla, jugs, Storybirds (1995)
Pekka Paikkari, pepper and salt set, Toi (1996)
Partanen/Kankkunen, cutlery, Artik (1997)
Carina Seth-Andersson, bowls from the Tools series (1998)
Annaleena Hakatie, bowls from the Relations Series (1998)
Erik Mangor, cheese slicer (1998)
Olof Soderholm, cheese slicer, Lamina (1995)
Ergonomi Design Gruppen, cutlery, Basic (1989)
Jorgen Moller, jugs and dishes from the Complet series (1993)
Louise Bach, cutlery, wooden dish, bowl (1998)

TEXTILES
Alvar Aalto, curtain fabric, Siena
Age Faith Ell, curtain fabric, Pling
Anna Sallander, curtain fabric, Rut
Vuoko Nurmesniemi, curtain fabric, Kirjeet Choponelle (1984)
Vuoko Nurmesniemi, curtain fabric, Lento (1984), Nopea (1984), Letters to Chopin (1984), Kuriiri (1984)
Margaret Adolfsdóttir, lasercut fabrics (1999)
Gudkaug Halldorsdóttri, fabric, La Marina (1998)

NORWAY

KUNSTINDUSTRIMUSEET I, OSLO

FURNITURE
Tias Eckhoff, chair, Bella (1995) manufactured by Smithco Line A/S
Gabriella Montaguti, adjustable table, Appresndista 2100 (1998) manufactured by Rexite SpA
Raul Barberi, hat and coat stand with incorporated umbrella stand, Doppiopetto 1090 (1992) manufactured by Rexite SpA
Jorge Pensi, outdoor chair, Toledo (1998) manufactured by Amat

LIGHTING
Phillip Deceuninck, light fitting, Techno, manufactured by Wever & Ducre.

TABLEWARE
Kati Tuominen-Niittyla, jugs, Story Birds (1997) manufactured by Arabia, Finland
Johan Verde, mugs and plates (1999) manufactured by Figgjo, Norway
Olav Joa, cup on a tray (1999) manufactured by Figgjo, Norway
Olav Joa, Inger Johanne Egeland, Bente Olsen, 3 cappuccino cups and 3 espresso cups (1999) manufactured by Figgjo, Norway
Andrea Branzi, water kettle, Mama-o (1992) manufactured by Officina SpA
Aldo Rossi, espresso machine, Ottagono (1993) manufactured by Officina Alessi
Giorgio Guroli and Francesca Scansetti, nutcracker, Per (1993) manufactured by Syn srl
Iron teapot, Arare (c. 1970), manufactured by Iwachu Foundly Co Ltd, Japan

Laco Pagac, glasses, LR2357 (1990), manufactured by Lr Crystal
Jozef Kolembus, glasses, LR 2376 (1995), manufactured by Lr Crystal

PRODUCTS
Ian Wilson, mixer, Sunbeam (1982) manufactured by Sunbeam Corp Ltd
Time Manager, Psion Organiser II (1983) manufactured by Psion Plc
Franceso Trabucco & Marcello Vecchi, wet and dry vacuum cleaner, Bidone Lavatutto (1986) manufactured by Alfatec/ Electrolux SpA
Shizuo Kakashino, walkman, TPS-L2 (1978) manufactured by Sony
Roy Tandberg, Davis Deskview (1988) manufactured by Davis AS, Norway

POLAND

NATIONAL MUSEUM, CENTRE OF MODERN DESIGN, WARSAW

TABLEWARE
Leszek Nowosielski, decorative plate, pottery (1970)
Bronislav Wolanin, mugs, pottery (1985)
Jan Siedlecki, glasses, Mono (1995)
Malgorzata Dajewska, crystal, Aquarius (1995)
Stanislaw Mozdzen and Anastazja Panek, coffee set, Opty (1996)
Piotr Talma, crystal glasses, Gothic (1997)

PRODUCTS
Andrzej Latos, rocking horse (1979)

SWEDEN

NATIONAL MUSEUM, STOCKHOLM

FURNITURE
Bjorn Dahlstrom, chair (1998), manufactured by CBI/Klara AB
Mattias Ljunggren, chair, Gute (1998) manufactured by Kallemo AB
Jonas Lindvall, sideboard (1998), manufactured by Angra AB
Ron Arad, rocking chair (1981)
Ross Lovegrove, chair, Figure of Eight (1993), manufactured by Cappellini SpA
Thomas Eriksson, cupboard, Cross (1992), manufactured by Cappellini SpA
Caroline Schlyter, chair, H (1989), manufactured by Forsnas AB
Thomas Sandell, chair, Akvarium (1998), manufactured by Garsnas
Bjorn Dahlstrom, armchair, BD5 (1997), manufactured by CBI/Klara
Jan Hellzen, table, Lack (1979), manufactured by IKEA
Thomas Sandell, bench, PS (1995), manufactured by IKEA
Marten Claesson/Eero Koivisto/Ola Rune, armchair, Berliner (1999), manufactured by Swedese Mobler AB
Thomas Sandell, chair, TC (1994), manufactured by Asplund
A&E Design, sofa, Dicken (1970), manufactured by Lammhults
A&E Design AB, stool, Stockholm II (1994)
Thomas Eriksson, stool, PS (1995), manufactured by IKEA
Erik Richter, armchair (1993), manufactured by Norrgavel
Jasper Morrison, chair, Ply Chair (1999), manufactured by Vitra AG
Ake Axelsson, armchairs (2000)

LIGHTING
Jonas Rooth, chandelier (1999)

TABLEWARE
Jonas Bohlin, vase, Liv (1997) manufactured by Reijmyre glasbruk
Maria Kariss, glass, Cascade (1999) manufactured by Rejmyre glasbruk
Maria Kariss, glass, Circle (1999) manufactured by Reijmyre glasbruk
Brita Flander, vase, Nice (1999) manufactured by Reijmyre glasbruk
Bente Hansen, vase (1999)
Carina Seth Andersson, vase 1999
Lars Fleming, bowl (1998), manufactured by Astelier Borgila AB
Erik Tidang, vase, Kotte (1999)
Ingegerd Raman, jug, Bellman (1991)
Ingegerd Raman, candlesticks (1992)
Ingegerd Raman, decanters (1995–7)
Tor Alex Erichsen, vase, Malstrom (1998)
Helen Krantz, vase (1998), manufactured by Orrefors Glasbruk
Per B Sundberg, glass (1998)
Marten Medbo, glass (1997), manufactured by Egerman Glasbruk
Torben Hardenberg, vase, Tarar (1999)
Per B. Sundberg, vase, Fabula (1998), manufactured by Orrefors Glasbruk
Marten Medbo, vase (1999), manufactured by Ajeto Glass
Maleene Mullertz, bowl (1999)
Thomas Sandell, pot, Louise (1998), manufactured by Garsnas
Asa Lindstrom, vase, Stories (1994), manufactured by Hackman Rorstrand AB
Annika Jarring, vase, Planglasvav no 4
Wolfgang Gessl, jug (1999)
Eva Bengtsson, jar, Kyrie Eleison (1999)
Signe Persson Melin, bowl, Galler (1999)
Ulrica Hydman-Vallien, vase, Mote I natten! (1990)
Ann Wahlstrom, vase, Sjosnacka
Ann Wahlstrom, vase, Hot Pink (2000), manufactured by Orrefors Glasbruk

TEXTILES
Inez Svensson, fabric print, Pampas (1998), manufactured by Kinnasand AB
Carl John Hane, fabric print, Tegelbacken (1998), manufactured by Kinnasand AB
Gunilla Lagerhem Ullberg, carpet, String (1998), manufactured by Kasthall
Pia Wallen, carpet, Krux (1993), manufactured by Asplund
Chiqui Mattson, quilt cover, Engelsk konfekt (1973) manufactured by Boras Wafveri
Irene Ajbaje, tapestry, Binary (1990)
Gunilla Lagerhem Ullberg, textile, Gunilla (1997)
Gunilla Lagerhem Ullberg, textile, Pling (1997)
Ines Svensson, fabric print, Randiga bananer (1986)
Kajsa af Petersens, textile, Badd (1996)
Britta Lincoln, broidery, Bygee (1999)
Lennart Rodhe, fabric print, Shiva (1976)

PRODUCTS
Jonathan Ive, computer, iMac (1998), manufactured by Apple Computer Inc.
Ergonomi Design Gruppen AB, body care programme, Beauty (1998)
Bjorn Dahlskog, cooking utensils, Hackman Tools (1998)
Designaktiebolaget Propeller, air cleaner, Blueair (1998)
Eva Abinger, mirror, Veritas (1998), manufactured by Simplicitas AB
Jan Olsen and Lars Samuelsen, bike, Itera (1978), manufactured by Itera AB
Ergonomi Design Gruppen, screwdrivers (1983), manufactured by Sandvik Bahco
Ergonomi Design Gruppen, adjustable spanners (1985), manufactured by Sandvik Bahco
Lars Eriksson/Bengt Pettersson, Kick-Sled, Monomed 89 (1995), manufactured by Vansbro Sparkfabrik
Morgan Ferm, walker, Tango (1999), manufactured by Etac AB, Stockholm
Ergonomi Design Gruppen, hammers (1993), manufactured by Hultafors AB
Bjorn Dahlstrom, bike, Z (1998)
Palm Inc/ IDEO, organizer, Palm Pilot V (1997)
Nokia Svenska AB, communicator, Nokia 9000 (1996)
Nokia Svenska AB, mobile telephone, NOKIA 8110 (1996)
Ergonomi Design Gruppen, hammers, T Block (1994)
Ergonomi Design Gruppen/Lillemor Jakobsen, baby carrier, Babybjorn 1980

RÖHSSKA MUSEET, GOTHENBURG

FURNITURE
Sergei V. Gerasimenko, chair (1993)
Carl-Arne Breger, child's bicycle chair (1975)
Ernst Billgren, cupboard, Stipendieskap

LIGHTING
El Ultimo Grito, lamp, Don't Run, We are Your Friends!
Droog Design, lamp, Soft
Ernesto Gismondi, lamp, e.Light (1999)

TABLEWARE
PO Landgren, spoon (1973)
PO Landgren, fork (1973)
PO Landgren, knife (1973)
Per B. Sundberg, glass vase, Fabula (1998)
Ingela Karlsson, earthenware bowl, Ribskal (1999)
Martti Rytkonen, glass vase Ett Monsteranfall I Blabarskogen (1999)
Sven-Erik Juhlin, pitcher (1970)
Jasper Morrison, salad set, Sim (1998)
Jasper Morrison, twin salad, Saladim (1998)
Pia Tornell, vase, Cirrus
Pia Tornell, porcelain bowls, Sinus (1994)
Asa Lindstrom, porcelain schnapps glass, Sofias (1999)
Asa Lindstrom, porcelain schnapps glass, Klaras (1999)
Asa Lindstrom, porcelain schnapps glass, Eriks (1999)
Asa Lindstrom, porcelain schnapps glass, Torstens (1999)
Handelsbolaget Porslinsfabriken I Gustavsberg, bowls, tray, Sushi
Carina Seth-Andersson, glass bowl, Relations
Sigurd Persson, silver pitcher (1999)
Bertel Gardberg, silver pitcher (1993)
Kati Tuominen-Niittyla, stoneware pitcher, Story-bird (1984)
Jane Reumert, bowl, Arabisk Musik (1999)
Ingegerd Raman, bowl, Strof (1990)
Ingegerd Raman, glasses, Strof (1990)
Ingegerd Raman, pitcher, Strof (1990)
Ingegerd Raman, pitcher, bowl, Mjolk Och Socker (1990)
Ingegerd Raman, glass bowl (1991)

PRODUCTS
Per B. Sundberg, porcelain radio (1999)
Frank Nuovo, cellular phone, Nokia 8810

SWITZERLAND

MUSEUM FÜR GESTALTUNG, ZÜRICH

FURNITURE
Klaus Vogt, rolltop cupboard, manufactured by Thut Mobel Moriken
Hanspeter Weidmann, shelving system, manufactured by Bigla AG Biglen
Lukas Buol and Marco Zund Basel, bed
Neue Werkstatt Winterthur, shelving unit

TEXTILES
Arthur David, Silk scarf

TABLEWARE
Luigi Tottoli, porcelain dinner service, manufactured by Pregassona

PRODUCTS
Mario Botta Lugano, wristwatch, Collection Pierre Junod, manufactured by Biel/Bienne
Stanley Tigerman, wristwatch, Collection Pierre Junod, manufactured by Biel/Bienne
Michael Graves, wristwatch, Collection Pierre Junod (1994), manufactured by Biel/Bienne
Michael Graves, wristwatch, Collection Pierre Junod (1996), manufactured by Biel/Bienne
Massimo Vignelli, wristwatch, Collection Pierre Junod, manufactured by Biel/Bienne
Shigeru Uchida, wristwatch, Collection Pierre Junod, manufactured by Biel/Bienne
Frederic Deddelly, wristwatch, Collection Pierre Junod, manufactured by Biel/Bienne
Ambroggio Pozzi, wristwatch, Collection Pierre Junod, manufactured by Biel/Bienne
Richard Meier, wristwatch, Collection Pierre Junod, manufactured by Biel/Bienne
Vito Noto Lugano, wristwatch, Collection Pierre Junod, manufactured by Biel/Bienne
Andre Ricard, wristwatch, Collection Pierre Junod, manufactured by Biel/Bienne
Xernex Zurich, Switzerland, wristwatch
Hannes Wettstein, wristwatches, manufactured by Ventura Wangen

UK

VICTORIA AND ALBERT MUSEUM, LONDON

FURNITURE
Barber Osgerby Associates, stool, Flight Stool (1998)
Barber Osgerby Associates, coffee table, Loop (1996)
Barber Osgerby Associates, shelf, Loop (1996)
Thomas Eriksson, cabinet, Red Cross Medicine Cabinet (1992), manufactured by Progetto

TABLEWARE
Hilary Roberts, porcelain jug (1999)
Hilary Roberts, porcelain bowl (1999)
Ursula Munch-Petersen, earthenware jug and lid, Ursula (1999)
Ursula Munch-Petersen, earthenware oval bowl, Ursula (1999)
Ursula Munch-Petersen, earthenware cup and plate, Ursula (1999)
Ole Jensen, carafe, glass, Ole (1995), manufactured by Royal Scandinavia A/S
Ole Jensen, porcelain carafe, Ole (1998/9), manufactured by Royal Scandinavia A/S
Ole Jensen, earthenware mug, Ole (1995), manufactured by Royal Scandinavia A/S
Ole Jensen, porcelain cup, Ole (1995), manufactured by Royal Scandinavia A/S
Ole Jensen, earthenware teapot and lid/spoon, Ole (1995), manufactured by Royal Scandinavia A/S
Ole Jensen, earthenware juice squeezer/pitcher, Ole (1995), manufactured by Royal Scandinavia A/S
Ole Jensen earthenware colander, Ole (1995), manufactured by Royal Scandinavia A/S
Peter Ting, bone china coffee can and saucer, Jewel Pattern (1997), manufactured by Thomas Goode & Co
Peter Ting, bone china tea cup and saucer, (1998), manufactured by Thomas Goode & Co
Peter Ting, bone china coffee cup and saucer (1998), manufactured by Thomas Goode & Co
Peter Ting, bone china sugar bowl and cover (1998), manufactured by Thomas Goode & Co
Peter Ting, bone china creamer (1998), manufactured by Thomas Goode & Co
Peter Ting, bone china coffee pot and cover (1998), manufactured by Thomas Goode & Co
Peter Ting, bone china 13" undecorated lay plate, 'Nebula' shape (1997), manufactured by Thomas Goode & Co
Peter Ting, bone china 10" dinner plate, 'Comma' pattern (1998), manufactured by Thomas Goode & Co
Peter Ting, bone china 8" dessert plate, 'Zen' pattern (1998), manufactured by Thomas Goode & Co
Peter Ting, bone china salad bowl, 'Metallic' pattern, 'Comet' shape (1998), manufactured by Thomas Goode & Co
Peter Ting, bone china dessert plate, 'Kaleidoscope' pattern (1997) manufactured by Thomas Goode & Co
Peter Ting, bone china soup plate, 'Kaleidoscope' pattern (1997), manufactured by Thomas Goode & Co
Peter Ting, bone china main course plate, 'Kaleidoscope' pattern (1997), manufactured by Thomas Goode & Co
Peter Ting, bone china presentation plate, 'Kaleidoscope' pattern (1997), manufactured by Thomas Goode & Co

TEXTILES
Caroline Broadhead, Block Dress (1999)
Freddie Robbins, gloves, Polly, Peter, Conrad and Giles (1999)
Bernhald Willhelm, gloves
Neisha Crossland, experimental silk woven scarf (2000)
Dorte Behn, woven linen scarves (1999)

USA

MUSEUM OF FINE ARTS, BOSTON

FURNITURE
Peter Shire, table, Brazil (1981), manufactured by Memphis
Peter Shire, table, Hollywood (1983)
Michele De Lucchi, table, Kristall, manufactured by Memphis
Ettore Sottsass, room divider, Carlton (1981), manufactured by Memphis
Aldo Cibic, console table, Belvedere Up & Up (1982), manufactured by Massa
Michael Graves, Stanhope Bed (1982), manufactured by Memphis
George James Sowden, chair, Palace Armchair (1983), manufactured by Memphis
Peter Shire, chair, Bel Air (1982)
Michael Graves, dressing table, Plaza (1981), manufactured by Memphis
Ettore Sottsass, bookcase, Factotum (1980), manufactured by Studio Alchimia
Ettore Sottsass, bookcase, Cargo (1979), manufactured by Studio Alchimia
Gerard Taylor, bookcase, Le Palme bookcase (1983), manufactured by Memphis
Ettore Sottsass, table, Le Strutture Tremano (1979), manufactured by Studio Alchimia

LIGHTING
Michele De Lucchi, lamp, Sinerpica (1979), manufactured by Studio Alchimicia
Michele De Lucchi, lamp, Oceanic (1981), manufactured by Memphis
Martine Bedin, lamp, Super (1981), manufactured by Memphis
Michele De Lucchi, lamp, Grand Floor Lamp (1983), manufactured by Memphis
Ettore Sottsass, floor lamp, Treetops (1981), manufactured by Memphis
Ettore Sottsass, floor lamp,Callimaca (1980), manufactured by Artemide

TABLEWARE
Ettore Sottsass, Stem Glass (1983), manufactured by Memphis
Marco Zanini, vase, Cassiopea (1982), manufactured by Memphis
Ettore Sottsass, Altair Glass (1982), manufactured by Memphis

Matteo Thun, Chad Teapot (1983), manufactured by Memphis
Matteo Thun, vase, Titicaca (1982), manufactured by Memphis
Matteo Thun, cup, Ladoga (1982), manufactured by Memphis
Matteo Thun, glass, Volga (1983), manufactured by Memphis
Ettore Sottsass, glass, Euphrates (1983), manufactured by Memphis
Ettore Sottsass, glass, Alcor (1983), manufactured by Memphis
Marco Zanini, covered bowl, Rigel (1983), manufactured by Toso Vetri d'arte
Michele De Lucchi, vase, Antares (1983), manufactured by Toso Vetri d'Arte
Ettore Sottsass, fruit dish, Murmansk (1982), manufactured by Rossi e Arcandi
Ettore Sottsass, dish, Sol Fruit (1982), manufactured by Toso Vetri d'Arte
Marco Zanini, Colorado Teapot (1983), manufactured by Memphis
Ettore Sottsass, Alioth Glass (1983), manufactured by Toso Vetri d'Arte
Marco Zanini, vase, Alpha Centauri (1982), manufactured by Toso Vetri d'Arte
Marco Zanini, fruit dish, Regolus (1983), manufactured by Toso Vetri d'Arte

TEXTILES
Alessandro Mendini, Oriented Carpet (1980)
Alessandro Mendini, Kandissone Tapestry (1980)
Nathalie du Pasquier, California Rug (1983)
Daniela Puppa, Explosant–Fixe Fabric Panel (1980), manufactured by Studio Alchimia
Fabric Panel (1980) manufactured by Studio Alchimia

PRODUCTS
Peter Shire, teapot, Anchorage (1983), manufactured by Rossi e Arcandi
George James Sowden, clock (1983)
Ettore Sottsass, Enorme telephone (1986), manufactured by Brondi Telephonia SpA
Matteo Thun, teapot (1982–3), manufactured by Alessi Quartet

BROOKLYN MUSEUM OF ART

FURNITURE
Andrée Putman, Teak Garden Collection Chair (1999), manufactured by Furniture Co
Konstantin Grcic, armchair, Cramer (1995), manufactured by Montina International
Konstantin Grcic, ES Shelf (1999), manufactured by Moorman
Mark Brazier-Jones, Lyre Console Table (1992)

LIGHTING
Andrée Putman, lamp, Compas dans l'Oeil (1999) manufactured by Baldinger Architectural Lighting, Inc.
Konstantin Grcic, lamp, Mayday Hanging Lamp, (1998) manufactured by Flos SpA

TABLEWARE
Achille Castiglione, tablespoon, fork, knife, dessert spoon, teaspoon, coffee spoon, Dry (1982–5), manufactured by Alessi
Don Gould, plate, Seaglass (2000), manufactured by Riverside Design Group Inc.
Don Gould, bowl, Seaglass, (2000), manufactured by Riverside Design Group Inc.
Don Gould, plate, The Architect Series (2000), manufactured by Riverside Design Group Inc.
Don Gould, plate, Origins (2000), manufactured by Riverside Design Group Inc.
Konstantin Grcic, decanter with top (1999), manufactured by Iitala.
Konstantin Grcic, ashtray with liner (1999), manufactured by Iitala
Konstantin Grcic, tray, (1999), manufactured by Iitala
Konstantin Grcic, goblet (1999), manufactured by Iitala
Konstantin Grcic , whiskey glass (1999), manufactured by Iitala
Konstantin Grcic, cocktail glass (1999), manufactured by Iitala
Konstantin Grcic, bowl (1999), manufactured by Iitala
Enzo Mari, fruit bowl, Adal (1968), manufactured by Alessi
Andrée Putman, dinner plate (1995), manufactured by Porcelaine Ancienne Manufacture Royale de Limoges
Andrée Putman, luncheon plate (1995), manufactured by Porcelaine Ancienne Manufacture Royale de Limoges
Andrée Putman, salad plate (1995), manufactured by Porcelaine Ancienne Manufacture Royale de Limoges
Andrée Putman, dessert plate (1995), manufactured by Porcelaine Ancienne Manufacture Royale de Limoges
Andrée Putman, cup and saucer (1995), manufactured by Porcelaine Ancienne Manufacture Royale de Limoges
Karim Rashid, 'Cube' Vessel (1999), manufactured by Totem Design Group
Karim Rashid, 'Soul' Vessel (1999), manufactured by Totem Design Group
Karim Rashid, 'Infinity' Vessel (1999), manufactured by Totem Design Group
Karim Rashid, 'Sectional' Vessel (1999), manufactured by Totem Design Group
Judy Simlow, plate, (2000), manufactured by Riverside Design Group
Ettore Sottsass, oil bottle, (1978), manufactured by Alessi
Ettore Sottsass, vinegar bottle (1978), manufactured by Alessi
Ettore Sottsass, salt and pepper shakers (1978), manufactured by Alessi

TEXTILES
Herman Toys Inc., textile sample (1989)

PRODUCTS
Andrea Branzi, kettle, Mama-o (1988), manufactured by Alessi
Frank Gehry, kettle, Pito (1988), manufactured by Alessi
Nathan George Horwitt, clock, Museum Wall Clock (1970), manufactured by Howard Miller Clock Company
Andrée Putman, fountain pen and rollerball pen (1999), manufactured by ACME Studios
Konstantin Grcic, Belle and Bon, egg cup and letterknife (1999), manufactured by Porzellen-Manufaktur Nyphemburg
Konstantin Grcic, Wanda (1997), manufactured by Wireworks Ltd
Richard Sapper, kettle with melodic whistle, (1983), manufactured by Alessi
Michael Graves, kettle with bird-shaped whistle (1985), manufactured by Alessi
Aldo Rossi, espresso coffee maker, La Conica (1982), manufactured by Alessi
Richard Sapper, espresso coffee maker (1979–92), manufactured by Alessi

THE CHICAGO ATHENAEUM

FURNITURE
Alfredo Walter Häberli and Christophe Marchand, Move-it Table Series (1997–9), manufactured by Davis Furniture Design
Allseating and Carbon Design, seating desk/workstation/stacking chair, Oss (1998), manufactured by Allseating Corporation
Christophe Hindermann, Orta Chair (1998–9), manufactured by Davis Furniture Industries Inc.
Dranger Design AB, SoftAir/ IKEA a.i.r chair, manufactured by News Design DFE A 3
Emil Lohrer and Marcus Schneider, chair, Sedus Y2K (1998–9), manufactured by Nienkamper
Forming Design, Sedus Smile Standing Seat (1998), manufactured by Sedus Stoll AG
Isao Hosoe Design, Please Indoor and Outdoor Chair (1998), manufactured by Segis SpA
Ohl Design, Tavola seating range (1998–9), manufactured by Wilkhahn Wilkening
Scot Laughton Studio, 'Juxta' storage/furniture, (1998–9), manufactured by Umbra Inc.
Stuart Basasches and Judy Hudson, shelving/table system, I-beam (1999), manufactured by Biproduct

LIGHTING
Isao Hosoe Design, HOI Lamps (1998), manufactured by Luxo Italiana SpA
Ross Lovegrove, Pod Lens Lamps, (1998–9), manufactured by LucePlan SpA
Alberto Meda and Poalo Rizzatto, Forebraccio Lamps, (1998–9), manufactured by LucePlan SpA
Ferdi Giardini, Blow Fan (1998–9), manufactured by LucePlan SpA
Ross Lovegrove, Solar Bud Walkway Lamps (1998–9), manufactured by LucePlan SpA
Agence Wilmotte, Spirit Track Light Fixture (1998), manufactured by Zumtobel Ag
Tobias Grau, TAA Ceiling and Suspension Lamp (1998), manufactured by Tobias Grau KG GmbH

TEXTILES
Suzanne Tick, Una upholstery fabric (1998–9), manufactured by Knoll Textiles
Suzanne Tick, Arras upholstery fabric (1998–9), manufactured by Knoll Textiles
Heinz Rontgen, Tre-de decorating fabric (1998–9), manufactured by Nya Nordiska Textiles GmbH
Heinz Rontgen, Scena Vista decorating fabric (1989), manufactured by Nya Nordiska Textiles GmbH
Frauke V. Jaruntowski, Design Concept Tinni and Tono decorating fabric, manufactured by Nya Nordiska, GmbH

PRODUCTS
ACCO Brands Inc., Swingline 415 heavy duty stapler (1997–8), manufactured by ACCO
ACCO Brands Inc., Swingline desktop stapler, (1998–9), manufactured by ACCO
ACCO Brands Inc., Swingline Worx '99 mini stapler (1998–9), manufactured by ACCO
Bruce Ancona and Louis Henry, EKCO Clip 'n Stay Clothespin (1998), manufactured by Ecko Housewares
Andersen Design Inc., Ergo Product Line (1998–9), manufactured by Household Products Inc.
Andersen Design, Inc, Ergo Mixer (1998–9), manufactured by Balck & Decker
Ashcraft Design, Loudspeaker Series (1997–9), manufactured by Infinity
Raul Barbieri, Carol Tayar, folding trolley, Tender (1998), manufactured by Rexite SpA
Blu Dot, 2D: 3D magazine rack (1998–9), manufactured by Blu Dot
Peter A Buchele, Buchele Basin (1999), manufactured by Edel-Stahl, Hard Austria
Antione and Philip Cohen, Laurastar Steamax steam station & professional iron (1998–9), manufactured by Divelit
ECCO Design Inc, Phenom Express hand-held personal computer (1998), manufactured by LG Eectronics
ECCO Design Inc., Boston Ultimate Standup Stapler (1998–9), manufactured by Hunt Corporation
ECCO Design, Colgate Grip'Ems children's toothbrush (1997–8), manufactured by Colgate Palmolive
Antti Eklund, Maus can opener & Teddy bottle opener collection, (1998–9), manufactured by Animal Design OY
Everex Color Freestyle Palm Personal Computer (1998–9), manufactured by Everex Systems Inc.
Susan Fabry and Mari Ando, OXO Good Grips soft handled cocktail forks and spreaders (1998), manufactured by OXO International
Fitch Inc., 'mimio' Portable Digital Meeting Assistant (1998–9), manufactured by Virtual Ink
Mark Gakewski, Aero ceiling fan, (1998–9), manufactured by Minka Aire
Michael Graves, 'Dreamscape' Series washbasin with vanity (1998–9), manufactured by Duravit AG
Chris Hardy, Cantilever Lantern (1998–9), manufactured by Design Ideas
Human Factors ID, OXO Good Grips and @ Hand Tools, (1998–9), manufactured by OXO International
Barbara Schmidt, tableware set, Update (1997–8), manufactured by Kahla/Thuringen Prozellian GmbH
Tim Kennedy, OXO Good Grips shrimp cleaner (1998–9), manufactured by OXO International
Tassilo von Grohlman, ashtray, Sancho Design ASH70/AL (1997–8), manufactured by Troika Boll GmbH
Tae Hyung Kim, Emergency Tool and Flash Light for the Car (1999)
LOEWE, Platinum Planus Consolette Widescreen Digital Color Television (1995–6), manufactured by LOEWE
Philip Bro Ludvigsen, coathanger, Hang on (1998), manufactured by Unique Interior
Peter Maly, Conmoto Chimney Set (1997), manufactured by Conmoto
Enzo Mari, Scrittura Desk Set (1999), manufactured by Rosenthal AG
D. Scott Miller, Tupperware Condimate (1997–8), manufactured by Tupperware Co
Fred Bould, Nambe Bridge 2000 Candelabra, (1998–9), manufactured by Nambé
Reiner Moll, Soho New York Bathroom Suite (1998–9), manufactured by Villeroy & Bosch
Reiner Moll, Ana Amalie Champagne Flute (1999), manufactured by Glas. Kunst. Lauscha GmbH
Tom Schonherr and Andreas Haug, Axor Terrano Single Lever Basin Mixer (1998–9), manufactured by Villeroy & Bosch
Roberto Pezzetta and Zanussi Industrial Design Centre, Oz Refrigerator (1994–8), manufactured by Electrolux Zanussi SpA
Roberto Pezzetta and Zanussi Industrial Design Centre, 'Input F1. 16 Electric Oven (1995–7), manufactured by Electrolux Zanussi SpA
Peter Sheehan and Lazhal Loughnane, Logitech Trackman Marble FX Mouse (1997–9), manufactured by Logitech Inc.
Peter Sheehan and Lazhal Loughnane, Logitech
Wingman Force Feedback Mouse (1999)
Sonja Pepcak, RC510 Rugged Remote Control, (1998–9), manufactured by Thomson Consumer Electronics
Karim Rashid, Suma, Tri, Jumbo,Bi and Rim Bowls (1998–9), manufactured by Umbra
Studio Brown, desk stapler, Isis 505, desk stapler, Rexite SpA
Philippe Starck, Axor Starck Two Handle Basin Mixer (1997–9), manufactured by Hansgrohne
Hermann Worner, coathanger and hook, Boogie (1996–7), manufactured by Weinbrecht & Partner GmbH
Werner Schlopp, Cerix Watch (1997–9), manufactured by RADO Watch Co Ltd
Umbra Design Group, Slip Shoe Horn (1999), manufactured by UMBRA Inc.
Nicole Zeller and Stewart Lee, Food Tool (1998–9), manufactured by Zelco Industries

ART INSTITUTE OF CHICAGO

FURNITURE
Henry P. Glass, Cricket Chair (1978), manufactured by Brown Jordan Co

DENVER ART MUSEUM

FURNITURE
Frank Gehry, table (1999)
Frank Gehry, chair, Skinny Beaver (1979–82), manufactured by Joel Stearns of New York City Editions
Frank Gehry, chaise longue, Bubbles (1987), manufactured by Urbangold
Michael Graves, coffee table (1986), manufactured by P. Gluck Woodworkers for Susan Hauserman
Michael Graves, Lounge Chair (1981), manufactured by Susan Hauserman
Toshiyuki Kita, Wink Chair (1980), manufactured by Cassina SpA
Shiro Kuramata, XI Side Chair (1983), manufactured by Ishimaru Co Ltd
Alessandro Mendini, Michael Thonet Chair (1978), manufactured by Studio Alchimia
Gaetano Pesce, Adult Chair (1999)
Gaetano Pesce, Childrens Chair (1999)
Borek Sipek, Gundrem Am Leineufer Chairs (1984), manufactured by Leitner Interior Design
Ettore Sottsass, Albert LeClerc, Masanori Umeda, Typing Chair (1972–3)
Studio 65, Capitelio Seating (1971), manufactured by Gufram Industria Arredamento srl
Ali Tayar, Plaza Screen (1999), manufactured by ICF Group
Robert Venturi, dining table, Urn (1984), manufactured by Knoll International
Shiro Kuramata, XI Side Chair (1983), manufactured by Ishimaru Co Ltd

LIGHTING
Ingo Maurer, lamp, Bulb (1980)
Gaetano Pesce, lamp (1999)
Gaetano Pesce, lamp, Friends (1999)
Ettore Sottsass, lamp, Halo Click 1 (1998–9), manufactured by Royal Philips Electronic

TABLEWARE
Bob Daenen of TEAM with Eric Herlow Design, Impressions Classic Bowl, Set 3 (1993), manufactured by Tupperware Corporation
Walter Gropius, Tac 1 Tea Service Model 1280 (1969), manufactured by Rosenthal Studio Line

TEXTILES
Jack Lenor Larsen, fabric, Nimbis (1980)
Gaetano Pesce, fabric, Motus (1968), manufactured by Expansion
Robert Venturi, Denise Scott Brown, Juniper Floral Gingham (1991), manufactured by Design Tex Inc.

PRODUCTS
Philippe Starck, colander, Max le Chinois (1990), manufactured by Alessi SpA
Robert Blaich, electric shaver, Philashave (1990), manufactured by Royal Philips Electronics
Robert Blaich, coffee maker, Café Gourmet (1989), Roller Radio (1986), manufactured by Royal Philips Electronics

LOS ANGELES COUNTY MUSEUM OF ART

FURNITURE
Sam Maloof, rocking chair 1950 (made 1997)

METROPOLITAN MUSEUM OF ART, NEW YORK

TABLEWARE
Yoichi Ohira, vase, Acqua alta di Venezia (1999)
Yasokichi Tokuda, porcelain vase (1997)

TEXTILES
Diane Itter, linen, Evolutionary phases (1979)
Hideo Yamakuchi, Jacquard brocade, Shisen (1997)

MUSEUM OF MODERN ART (MoMA), NEW YORK

FURNITURE

Philippe Starck, W.W. Stool (1990), manufactured by Vitra AG
Philippe Starck, Royalton Bar Stool (1998), manufactured by XO
Philippe Starck, Miss Trip Chair (1996), manufactured by Kartell
Philippe Starck, Miss Coco Folding Chair (1998), manufactured by Cassina
Philippe Starck, Lola Mundo Chair (1988), manufactured by Driade
Philippe Starck, Louis XX Chair (1991), manufactured by Vitra
Philippe Starck, Dadada Rocking Stool, manufactured by OWO
Fernando Campana, Humberto Campana, Ottoman (1998), manufactured by Campana Objetos Ltda
Fernando Campana, Humberto Campana, Vermehla Chair (1993), manufactured by Edra Mazzei
Fernando Campana, Humberto Campana, Cone Chair (1997), manufactured by Edra Mazzei
Fernando Campana, Humberto Campana, 'Inflating Table' (1996) Campana Objetos Ltda
Philippe Starck, Aleph Stool (1992), manufactured by Driade
Philippe Starck, Bubu 1er Stool (1991), manufactured by OWO
Philippe Starck, La Marie Folding Chair (1988), manufactured by Kartell
Philippe Starck, Peninsula Chair (1995), manufactured by XO
Ross Lovegrove, Magic Chair (1997), manufactured by Fasem International

LIGHTING

Gugliemo Berchicci, floor lamp, Loto (1997), manufactured by Kundalini srl
Stefano Giovannoni, lamp, Big Switch (1996), manufactured by Segno srl
Konstantin Grcic, lamp, May Day (1998), manufactured by Flos
Harry Koskinen, lamp, Block (1996), manufactured by Design House Stockholm AB
Ingo Maurer, lamp, Wo bist Du, Edison…? (1997), manufactured by Ingo Maurer GmbH
Ingo Maurer, Zettel'z Hanging Lamp (1997), manufactured by Ingo Maurer GmbH
Ingo Maurer, table lamp, Los Minimalos Dos (1994), manufactured by Ingo Maurer GmbH
Ingo Maurer, wall lamp, Lucellino (1992), manufactured by Ingo Maurer GmbH
Philippe Starck, table lamp, Miss Sissi (1991), manufactured by Flos
Philippe Starck, Luci Fair Wall Sconce (1989), manufactured by Flos

TABLEWARE

Mikael Bjornstjerna, cheese knife, Oval Steel (1976), manufactured by Boda Nova
Philippe Starck, vase, Petite Etrangete contre un mur (1988), manufactured by Daum
Philippe Starck, toothpick holder set, Mister Cleen (1996), manufactured by Officina Alessi
Philippe Starck, cheese grater, Mister Meumeu (1992), manufactured by Officina Alessi

TEXTILES

Koichi Yoshimura, Iridescent Satin (1994), manufactured by S. Yoshimura Co Ltd
Koichi Yoshimura and Reiko Sudo, Blue Mirror Cloth with Wrinkles (1995)
Reiko Sudo, patched paper, polyester and mino washi paper, (1997), manufactured by Nuno Corporation
Reiko Sudo, Shutter, nylon (1997), manufactured by Nuno Corporation
Urase Co Ltd, Harmony (1997), manufactured by Urase Co Ltd
Keiji Otani, Brickyard, nylon and polyurethane (1997), manufactured by Nuno Corporation
Sakase Adtech, Triaxial Fabric (1991), manufactured by Sakase Adtech Co Ltd
Reiko Sudo, Moss Temple (1997), manufactured by Nuno Corporation
Akihiro Kaneko, polyester and washi paper, Yuragi (Fluctuation) (1997), manufactured by Kaneko Orimoni Co Ltd
Yoshihiro Kimura, nylon, polyurethane, polyester, rayon, Pedocal (1996), manufactured by Kimura Semko Co Ltd
Osamu Mita, Washi and Wool (1997), manufactured by Mitasho Co Ltd

PRODUCTS

Philippe Starck, Toothbrush and Toothbrush Holder, (1989), manufactured by Fluocaril
Philippe Starck, Walter Wayle Wall Clock (1989), manufactured by Officina Alessi
Philippe Starck, Poe Radio (1996), manufactured by Officina Alessi
Philippe Starck, ashtray, Ray Hollis (1986), manufactured by XO
Philippe Starck. table mirror, Miss Donna (1987), manufactured by OWO
Philippe Starck, colander, Max le Chinois (1987), manufactured by Officina Alessi
Philippe Starck, drawer handle, Mimi Bayou (1987), manufactured by OWO
Philippe Starck, lemon squeezer, Juicy Salif (1990), manufactured by Officina Alessi
Philippe Starck, knife, Laguiole (1986), manufactured by OWO
Philippe Starck, Hot Fredo Thermos (1993), manufactured by Alfi
Philippe Starck, portable television, Jim Nature (1984), manufactured by Thomson for Saba
Philippe Starck, ashtray, Joe Ccatus (1990), manufactured by Officina Alessi
Philippe Starck, Dr Skud Fly Swatter (1998), manufactured by Officina Alessi
Philippe Starck, Excalibur Toilet Brush (1993), manufactured by Heller USA
Philippe Starck, Hot Bertaa Tea Kettle (1990–91), manufactured by Officina Alessi
Philippe Starck, toothbrush, Dr Kiss (1996), manufactured by Officina Alessi
Philippe Starck, Dr Kiss Toothbrush (1996), manufactured by Kartell
Philippe Starck, cake server, Ceci n'est pas une truelle (1996), manufactured by Officina Alessi
Philippe Starck, Chab Wellington Coat Hook (1987), manufactured by OWO
Philippe Starck, Coo Coo Radio Alarm Clock (1996), manufactured by Officina Alessi
Philippe Starck. radio, Moosk (1996), manufactured by Officina Alessi
Swatch, GB 001 Watch (1983), manufactured by Swatch AG
Swatch, GK100 Watch (1983), manufactured by Swatch AG
Swatch, SDK 100 Skin Jelly, manufactured by Swatch AG

PHILADELPHIA MUSEUM OF ART, PENNSYLVANIA

FURNITURE

Beata Bar and Harmut Knell, chair, Migros (1997)
Max Ernst, bed and screen, Bed-Cage (1974)
Diego Giacometti, pair of armchairs (1973)
Diego Giacometti, pair of tables, Carcass (1970)
Diego Giacometti, table (1973)
Robert Wilson, pair of chairs, Queen Victoria (1974)
Maya Lin, Knoll Inc., seats, Stones (1998)
Karim Rashid, table, Aura (1990) manufactured by Zeritalia
Robert Venturi, Paola Navone, Michael Womack, chair, Gothic Revival (1979–1984), manufactured by Knoll International

LIGHTING

Ingo Maurer, lamp, Mozzkito (1996)

TABLEWARE

Karim Rashid, sugar bowl and creamer, Jimmy (1995) manufactured by Nambe
Karim Rashid, Rimbowl (1998) manufactured by Umbra Ltd

TEXTILES

Venturi, Scott Brown and Associates, fabric, Dots (1991) manufactured by Fabric Design Tex

PRODUCTS

Apple Industrial Design Team, computer, iMac (1998)
Karim Rashid, wall clock (1960) manufactured by Gyro Umbra Ltd
Karim Rashid, Garbo Can (1999) manufactured by Gyro Umbra Ltd
Karim Rashid, Garbino Can (1996) manufactured by Gyro Umbra Ltd

THE COOPER HEWITT NATIONAL DESIGN MUSEUM, NEW YORK

FURNITURE

Olaf Von Bohr, bookshelf/desk (1969–75) manufactured by Kartell
Dakota Jackson, Library Chair (1991)

TABLEWARE

Gerald Gulotta, porcelain dinner service, Chromatics (1970), manufactured by Arzberg Porzellanfabrik
Gerald Gulotta, 9 piece glass table setting, Chromatics (1970), manufactured by Cristais de Alobaca
Gerald Gulotta, 5 piece flatware setting, Chromatics (1970), manufactured by La Industrial Mondragonesa
Gerald Gulotta, 5 piece flatware setting, Iona (1979), manufactured by Yamazaki Kinzoku Kogyo
Gerald Gulotta, 5 piece flatware setting, Espana (1968–71), manufactured by La Industrial Mondragonesa
Gerald Gulotta, set of 6 cordial glasses, Atlantis (1970), manufactured by Cristais de Alcobaça

PRODUCTS

Philippe Starck, radio, Oye Oye (1994), manufactured by SABA
Pocket camera, Olympus O-Product (1988), manufactured by Olympus
Al Nagele and Leon Soren, telephone, Star Tac™ (1996), manufactured by Motorola Inc.

Suppliers

Agape srl, Via Ploner 2, 46038 Mantova, Italy.
T. (0)0376 371 738 F. (0)0376 374 213
E. agape@interbusiness.it

Werner Aisslinger, 33 Leibnizstrasse, Berlin, Germany.
T. (0)362 99 58 52 F. (0)30 31 50 54 01

Alcatel Telecom, 32 Avenue Kleber, 92707 Colombes, Cedex, France.
T. (0)1 55 66 7961 F. (0)1 55 66 7495
E. Anne.Bigand@alcatel.fr

Aiias srl, Via dei Videtti 2, I-24064 Grumello del Monte, Bergamo, Italy.
T. (0)035 44 22 511 F. (0)035 44 22 50
E. info@aliasdesign.it

Alessi SpA, Via Privata Aiessi 6, Crusinallo, Verbania, VB, Italy.
T. (0)323 86 86 11 F. (0)323 86 61 32
E. pub@alessi.it

Harry Allen & Associates, 207 Avenue A, New York, NY 10009, USA.
T. (0)212 529 7239 F. (0)212 529 7982
E. hasallen@earthlink.net

Antonangeli illuminazione srl, Via Volterra 10, 20052 Monza, (MI) Italy. T. (0)039 2720552
F. (0)039 2720796 E. antoill@tin.it

Anthologie Quartett, Schlosshuennefeld, 49152 Bad Essen, Germany.
T. (0)5472 940 90 F. (0)5472 940 940

Apple Computer Inc., 20730 Valley Green Drive, Cupertino 95014, California, USA
T. (0)650 728 0530 F. (0)415 495 0251

Ron Arad Associates Ltd, 62 Chalk Farm Road, London NW1 8AN, UK.
T. (0)207 284 4963/5 F. (0)207 279 0499
E. ronarad@mail.pro-net.co.uk

Arflex International SpA, Via Don R. Bereta 12, Guissano (MI) 20034 Italy.
T. (0)362 85343 F. (0)362 853080
E. info@arflex.it

Arnolfi di Cambio, Compagnia Italiana del Cristallo srl, 53034 Colle di Val d'Elsa, Siena Loc. Pian del'Olmino, Italy.
T. (0)577 9282 79 F. (0)577 92 96 47
E. dicambo@cyber.dada.it

Artelano, 4 Rue Schoelcher, 75014 Paris, France.
T. (0)1 43 22 74 91 F. (0)122 84 49
E. artelano@aol.com

Artemide SpA, Via Canova 34, Milan 20145, Italy.
T. (0)2 34 96 11 1 F. (0)2 34 53 82 11
E. pr@artemide.com

Arzenal, Valentinska 11, 11001 Prague 1, Czech Republic. T. (0)2 248 14099 F. (0)2 2481 0722
E. galerie@arzenal.cz

Asplund, 31 Sibyllegatan, Stockholm 11442, Sweden. T. (0)8 662 52 84 F. (0)8 662 38 35

Ateiier Pelcl, Farskeho 8, 17000 Prague, Czech Republic. T. (0)02 878869 E. atelier@pelcl.cz

Atrox GmbH, Seeblick 1, Cham, CH-6330, Switzerland. T. (0)41 785 0410 F. (0)41 785 0419
E. atrox@datazug.ch

Authentics, Max Eyth Strasse 30, Holzgerlingen D-71088, Germany.
T. (0)7031 68050 F. (0)7031 6805 99

Francois Azambourg, 19 rue Morice, 92110 Clichy la Garenne, France.
T. (0)47 30 17 07 F. (0)1 47 30 17 07

Azumi, 953 Finchley Road, London NW11 7PE, UK.
T. (0)208 731 9057 F. (0)208 731 9057

Emmanuel Babled, Via Segantini 71 Milan 20143, Italy. T. (0)2 58 11 11 19

Valter Bahcivanji, R. Alvaro Annes 79, 05421 010 Sao Paulo, Brazil.
T. (0)5511 2123168 F. (0)5511 69124841

Gijs Bakker Design, Keizergracht 518, 1017 EK Amsterdam. T. (0)20 638 2986 F. (0)20 6388828
E. gbakker@xs4all.nl

B & C Design, Genferstrasse 33, CH-8002 Zurich, Switzerland.
T. (0)01 201 27 47 F. (0)01 201 27 48
E. bc-design@pop.agri.ch

Baleri Italia, Via F. Cavalotti 8, Milan 20122, Italy.
T. (0)2 76 01 46 72 F. (0)2 76 01 44 19
E. info@baleri-italia.it

Ralph Ball, 177 Waller Road, London SE14 5LX UK.

Belux AG, Bremgarterstrasse 109, CH-5610 Wohlen, Switzerland.
T. (0)56 618 73 73 F. (0)56 618 73 27
E. belux@belux.ch

Bent Krogh A/S, Gronlandsvej 5, Postboks 520, DK-8660 Skanderborg, Denmark.
T. (0)86 52 09 22 F. (0)86 52 36 98
E. bk@bent-krogh.dk

Bernstrand & Co, Skanegatan 51, 11637 Stockholm.
T. (0)070 4224 05 00 F. (0)08 641 91 20
E. mail @bernstrand.nu

Marc Berthier, Design Plan Studio, 141 BD St Michel, 75005 Paris, France.
T. (0)1 43 26 49 97 F. (0)1 43 26 54 64
E. dpstudio@wanadoo.fr

Jurgen Bey, Passeralstraat 44A 3023 ZD Rotterdam, The Netherlands.
T. (0)10 425 8792 F. (0)10 425 9437
E. bey@luna.nl

Bius-Mary Little and Peter Wheeler, 120 Battersea Business Design Centre, 103 Lavender Hill, London SW11 5AL, UK.
T. (0)207 924 7724 F. (0)207 924 6524

Marc Boase, 33 Lower Market Street, Brighton BN3 1AT, UK. T/F. (0)1273 328 603
E. marcboase@pavilion.co.uk

Bodum AG, Kantonstrasse 100, 6234 Triengen, Switzerland. T. (0)41 935 45 00 F. (0)41 935 45 80

Boffi SpA, Via Oberdan 70, Lentate sul Seveso 20030, Milan, Italy.
T. (0)362 5341 F. (0)362 56 50 05
E. boffimarket@boffi.it

Bosch Telecom GmbH, John F. Kennedy Str. 43-53, D 38228 Salzgitter, Germany.

Botium, Bella Center, Center Blv. 5, 2300 CPH, Denmark. T. (0)32 51 69 65 F. (0)32 51 33 45

Boum Design, 527 East 6th Street 5w, New York, NY 10009, USA.
T. (0)212 254 1070 F. (0)212 431 6121

Ronan Bouroullec, 15 Rue des Ursulines, Saint Denis 93200, Paris, France. T. (0)1 48 20 36 60

Bower GmbH, 1 Mettinger Strasse, Neuenkirchen D-49586, Germany.
T. (0)5465 9292 0 F. (0)5464 9292 15

Box Design AB, Repslagargatan 17b, 1tr, 11846 Stockholm, Sweden.
T. (0)46 6 6401212 F. (0)46 6 6401216

Britefuture, Via longhi 6, 20137 Milan, Italy.
T/F (0)273 80327 E. britefuture@tiscalnet.it

Build, 4-2-11 Jingu Mae, Shibuya-Ku, 150 0001 Tokyo, Japan.
T. (0)81 3 3405 1211 F. (0)81 3 3401 9798

Bulthaup GmbH & Co, D-84153 Aich, Germany.
T. (0)49 1802 212534

Büro für form, Hans Sachs-Strasse 12, 80469 Munich, Germany.
T. (0)89 2 949 000 F. (0)89 26 949 002
E. buerofuerform@metronet.de

Sigi Bussinger, Hellabrunnerstrasse 30, Munich 81543, Germany.

Bute, c/o Caro Communications, 1st Floor, 49–59 Old Street, London EC1V 9HX, UK.
T. (0)207 251 9112 F. (0)171 490 5757
E. pr@carocom.demon.co.uk

Byra Interior Objects, Lessingstrasse 12, 65189 Wiesbaden, Germany.
T. (0)611 37 32 46 F. (0)611 37 51 31

Canon Inc, 30–2 Shimaruko 3-chome, 146-8501, Ohta-ku, Japan. T. (0)7 58 2111 F. (0)3 5482 9711
E. richard@cpur.canon.co.jp

Casas M. sl, Polignono Santa Rita, PO Box 1.333, E-08755 Castellbisbal, Barcelona, Spain.
T. (0)93772 4600 F. (0)93772 2130
E. casas@casas.net

Chiara Cantono, Via Malpighi 3, 20129 Milan, Italy. T. (0)2 29518792 F. (0)2 29518189

Cappellini SpA/Units srl, Via Marconi 35, Arosio (Como) 22060, Italy.
T. (0)31 75 91 11 F. (0)31 76 33 22

Cassina SpA, Via L Bisnelli 1, Meda/Milan, Italy.
T. (0)362 372 1 F. (0)362 34 22 46/34 09 59
E. info@cassina.it

ClassiCon GmbH, Perchtinger Strasse 8, Munich 81379, Germany.
T. (0)89 7 89 99 96 F. (0)89 7 80 99 96
E. info@classicon.com

Ciatta a Tavola, 50010 Badia a Settimo, Firenze, Italy. T. (0)55 7310817 F. (0)55 73310827
E. maciatti@tin.it

Cibic & Partners, Via Varese 18, Ingresso Viale F. Crispi 5 20121 Milan, Italy.
T.(0)2 657 1122 F. (0)2 290 601 41 E. cibic@tin.it

ClassiCon GmbH, Perchtinger Strasse 8, Munich 81379, Germany.
T. (0)89 7 89 99 96 F. (0)89 7 80 99 96

Codice 31, Via le Regina Giovanna 26, Milan 20129, Italy.
T. (0)2 29 51 77 22 F. (0)2 66 98 33 87

Covo srl, Via Degu Olmetti 3/b, 00060 Formello, (Rome) Italy. T. (0)06 90400311
F. (0)06 90409175 E. mail@covo.it

De Padova srl, Corso Venezia 14, Milan 20121, Milan, Italy. T. (0)2 77 720 1 F. (0)2 78 40 82

D4 Industrial Design, Nieuevaart 128, 1018 ZM Amsterdam, The Netherlands. T. (0)20 7760018
F. (0)20 7760019

Design Tech, Zeppelinstrassse 53, 72119 Ammerbuch, Germany.
T. (0)7073 918 90 F. (0)7073 918 917
E. info@designtechschmid.de

Design 3 Produktdesign, Schaarsteinwegsbrucke 2, D-20459 Hamburg, Germany.
T. (0)40 378 793 00 F. (0)40 378 793 09

De Lucchi srl (Produzione Privata), Via Pallavicino 31, Milan 20145, Milan.
T. (0)2 43 00 81 F. (0)2 43 11 82 22

Design Plan Studio, 141 Boulevard St Michel, 75005 Paris, France.
T. (0)43 26 49 97 F. (0)43 26 54 62
E. dpstudio@wanadoo.fr

Disegni, Via Gaudenzio Ferrari 5, 20123 Milan, Italy. T. (0)2 58114412 F. (0)2 58114412
E. disegni@disegni.com

dix heures dix, 127 Avenue Daumesnil, 75012 Paris, France.
T. (0)01 43 40 74 60 F. (0)01 43 40 74 85
E. DIXHEURESDIX@wanadoo.fr

Droog Design, Sarphatikade 11, 1017 WV Amsterdam, The Netherlands.
T. (0)20 62 69 809 F. (0)20 63 88 828
E. droog@euronet.nl

Dumoffice, Tollenstraat 60, 1053 RW Amsterdam, The Netherlands.
T. (0)020 489 0104 F. (0)020 4890104
E. dumoffice@planet.nl

Eastman Kodak Company, 901 Elmgrove Road, Rochester, 14653-5209 New York, USA.
T. (0)716 726 4413 F. (0)716 726 9255
E. FRANCIS.SKOP@KODAK.COM

Edra SpA, Via Livornese Est 106, Perignano 56030, Italy. T. (0)587 61 66 60 F. (0)587 61 75 00

El Ultimo Grito, 26 Northfield House, Frensham Street, London SE15 6TL, UK.
T. (0)207 732 6614 F. (0)207 277 6
E. grito@BTinternet.com

Eva Denmark A/S, 59 Højnæsvej, Rodovre, DK-2610, Denmark. T. (0)36 73 20 60 F. (0)36 70 74 11
E. mail@evadenmark.com

Khodi Feiz, Nieuwe prinsengracht 122 hs, 1018 VZ Amsterdam, The Netherlands.
T. (0)20 42230929 F. (0)20 4230928
E. feizdesignstudio@wxs.nl

Flos, Via Angelo Faini 2, 25073 Bovezzo (Bs), Italy.
T. (0)3024381 F. (0)302 711578

Flou SpA, Via Cadorna 12, 20036 Meda, (MI) Italy.
T. (0)362 37 31 F. (0)0362 748801
E. infoflou@flou.it

FontanaArte, Alzaia Trieste 49, Corsico 20094, Milan, Italy. T. (0)2 45 12 330

Fredericia Furniture AS, Treldevej 183, Fredericia 7000, Denmark. T. (0)7592 3344 F. (0)7592 3876

Galerie Kreo, 11 Rue Louise Weiss, 75013 Paris, France.
T. (0)153601842 F. (0)153601758
E. kreogal@club-internet.fr

Garcia Garay Illuminación Diseño, San Antonio 13, Sta Coloma. De Gramenet, 08923, Barcelona, Spain. F. (0)93 386 23 72

Gervasoni SpA, 33050 Pavia di Udine, Italy.
T. (0)432 656611 F. (0)432 656612

Christian Ghion, 156 Rue Oberkamph, Paris 7501, France. T. (0)1 49 29 06 90 F. (0)1 49 29 06 89

Giga, 23 Vinohradska, 12000 Prague 2, Czech Republic. T. (0)02 22250662 F. (0)02 2225 4436
E. gigaline@giga.cz

Natanel Gluska. Renggerstrasse 85, CH-8038 Zurich, Switzerland. T. (0)1 48 30 366

Tobias Grau KG GmbH & Co, Siemensstr. 35b, D-25462 Rellingen. T. (0)4101 37 00
F. (0)4101 370 1000 E. info@tobiasgrau.com

Gitta Gschwendtner, Unit 25, Oxo Tower, Barge House Street, London SE1 9PH, UK.
T (0)207 928 0143

Dögg Gudmundsdóttir, Lundtoftegade 26 1 th, 2200 Copenhagen, Denmark.
T. (0)35 85 36 35

Heller Inc., USA, 41 Madison Avenue, New York, NY 10010, USA.
T. (0)212 685 4204 F. (0)212 685 4204

Horm, Via Crocera di Corva 25, 33082 Azzano Decimo, Italy.
T. (0)0434 640733 F. (0)04334 640735

Richard Hutten, 52 Marconistraat, Rotterdam 3029, The Netherlands.

IDEO Europe, White Bear Yard, 144a Clerkenwell Road, London EC1R 5DF, UK.
T. (0)20 7713 2600 F. (0)20 7713 2601

Takashi Ifuji, Via Gusti 5, 20154 Milan, Italy.
T/F (0)2 31 6278 E. ifuji@micronet.it

IKEA of Sweden, Box 702, Almhult 34381, Smaland, Sweden. T. (0)47681012 F. (0)476 15123

Interlübke Gebr. Lübke GmbH & Co KG, Ringstrabe 145, 33378 Rheda Wiedenbrucke, Germany. T. (0)05242 12 232 F. (0)05242 12 311

Iridium, 709 Hyundaigolden tel 102, Kwang Jang Dong, Seoul, Korea. T. (0)82 2 34376525 F. (0)82 2 3437 6524

Dakota Jackson Inc., 42–24 Orchard Street, Fifth Floor, LIC, New York, NY 11101, USA.
T. (0)718 7886 8600 F. (0)718 706 7718

Jam design & communications Ltd, 2nd floor, 1 Goodsway, London NW1 1UR, UK.
T. (0)20 7278 5567 F. (0)20 7278 3263
E. all@jamdesign.co.uk

Claudy Jongstra, Openhartsteeg 1, 1017 BD Amsterdam, The Netherlands.
T. (0)31 20 42 84 230 F. (0)31 20 62 09 673
E. NTDH@wxs.nl

Kartell SpA, Via della Industrie 1, Noviglio (MI) 20082, Italy. T. (0)2 90 01 21 F. (0)2 90 53 316

Yehudit Katz, 11 Natan Hechachan, Tel Aviv 63413, Israel.

Knoll International Ltd, 1 Lindsey Street, East Market, Smithfield, London EC1A 9PQ, UK.
T. (0)207 236 6655 F. (0)207 248 1744

Kreon NV, Frankriojklei 112, 2000 Autlerpen, Belgium. T. (0)2 231 24 22 F. (0)3 231 88 96
E. mailbox@kreon.be

Lammhults Möbel AB, PO Box 26, Lammhult 360 30, Sweden.
T. (0)472 2695 00 F. (0)472 2605 70
E. info@lammhults.se

Danny Lane, 19 Hythe Road, London NW10 6RT, UK. T. (0)208 968 3399 F. (0)208 968 6289

Lemongras Production, Thalkirchnerstrasse 59, D-80337 Munich, Bavaria, Germany.
T. (0)089 7205 9911 F. (0)89 7205 9912
E. people@lemongras.it

Leonardo Glass, Industriegebiet Herste, 33014 Bad Driburg, Germany.
T. (0)5253/86 0 F. (0)5253/ 86 325

Lerche, Pet. Dan. Baus. D-42653 Solingen, Germany. T. (0)212 5 97 97 F. (0)212 59 25 06
E. info@lerche-solingen.de

Leucos SpA, Via Treviso 77, 30037 Scorze (VE), Italy. T. (0)041 5859111 F. (0)041 447598

Ligne Roset, Serrieres de Briord, Briord 01470, France. T. (0)4 74 36 17 00 F. (0)4 74 36 16 95

K.C. Lo, 31 Finsbury Park Road, London N4 2JY, UK. T. (0)7767 372 907 F. (0)207 503 8049
E. kc@netmatters.co.uk

Loom, Goodbert Reisenthal, Justus vin Liebig, Strasse, 3, D-86899 Landsberg, Germany.
T. (0)81 91/91 94 260 F. (0)81 91 94 129
E. info@lloydloom.de

Lumina Italia srl, Via Casorezzo 63, 20010 Milan, Italy. T. (0)02 9037521 F. (0)02 90376655
E. Market@lumina.it

Massimo Lunardon, Via Mottarello 6, 36060 Molvena (VI), Italy.
T. (0)0424 708329 F. (0)0424 708329

W. Lusty & Sons Ltd, Lloyd Loom Furniture, Hoo Lane, Chipping Campden, Glos., GL55 6AU, UK.
T. (0)1386 841333 F. (0)1386 841322

Luxo Italiana SpA, 1 Via Delle More, 24030 Presezzo (BG), Italy.
T. (0)035 603511 F. (0)035 464817
E. office@luxo.it

Ane Lykke, Ole Suhrsgade 8 3tv, Copenhagen 1354, Denmark. T. (0)3314 8064 F. (0)3314 0864

Magis srl, Via Magnadola 15, Motta di Livenza (Treviso) 31045, Italy.
T. (0)42276 87 42/3 F. (0)422 76 63 95

Mandarina Duck Plastimoda, 13 Via Minghetti, 40057 Cadriano Di Granorola, Italy.
T. (0)51 764 506 F. (0)51 602 0508
E. pcstile@mandarinaduck.it

Matsushita Electric Industrial Co Ltd, 2-1-61 Shiromi, 540-6214 Chuo-ku, Osaka, Japan.
T. (0)6 6949 2043 F. (0)6 6947 5606

Marset Illuminacion SA, Alfonso XII 429, 08918 Badalona, Spain.
T. (0)93 4600107 F. (0)93 4601089

Martela OYJ, Strombereinte, 11380 Helsinki, Finland. T. (0)10 37550 F. (0)10 345 5744

Mauser Office GmbH, 34513 Waldeck, Germany.
T. (0)05623 / 581-0 F. (0)05623 581 208

Meccano, Domein 4, 1261 JP Blaricum, The Netherlands T. (0)355395450 F. (0)355395456
E. info@meccano.nl

Memphis, Via Olivetti 9, Milan 20010, Italy.
T (0)2 93 29 06 83 F. (0)2 93 59 12 02
E. memphis.milano@ tiscalinet.it

Merloni Elettrodomestici, Viale Aristide Merloni 47, 60040 Fabriano (AN) Italy. T. (0)732 6611
F. (0)732 662954 E. digital@merloni.com

Mimolimit, Jakubska 2, 11000 Prague 2, Czech Republic. T. (0)2 24 81 42 76 F. (0)2 24 81 42 76
E. architect@architect.cz

Mobileffe, Via Ozanam 4, 20031 Cesano Maderno, (MI) Italy. T. (0)0362 52941
F. (0)0362 502212 E. mobileeffe@uli.it

Modular Lighting Instruments, Rumbeeksesteenweg 258, 8800 Roeslare, Belgium. T. (0)51 25 27 25 F. (0)51 25 27 88
E. edouard@supermodular.com

Molteni SpA, Via Rossini 50, 20034 Ginssano, Italy. T. (0)0362 851 334 F. (0)0362 35 51 70

Monotub Industries, 90 Long Acre, London, WC2E 9RZ, UK.
T. (0)20 7917 1863 F. (0)20 7917 1883
E. brian.austin@monotub.co.uk

Montina International srl, Via Communale del Rovere 13/15, 33048 S. Giovanni al Natisone, Udine, Italy.
T. (0)0432 75 60 81 F. (0)0432 75 60 36

Moormann Möbel GmbH, 2 An der Festhalle, Aschau i. Ch D-83229, Germany.
T. (0)8052 4001 F. (0)8052 4393
E. info@moormann.de

Moroso SpA, Via Nazionale 60, Cavalicco do Tavagnacco 1-33010 Udine, Italy.
T. (0)432 577 11 F. (0)432 570 761
E. info@moroso.it

Murano Due, Via delle Industrie 16, 30030 Salzano, Italy.
T. (0)041 5740292 F. (0)041 5744070
E. m2export@muranodue.com

NAC Sound srl, Via Boncompagni, Rome, 7900187, Italy. T. (0)6 44 55 730

Gabriela Nahlikova, 12 u Uranie, 17000 Prague, Czech Republic. T. (0)00420 2876685
E. navassdva@hotmail.com

Nava Design SpA, Martin Lutero 5, Milan 20126, Italy. T. (0)2 25 70 251 F. (0)2 26 30 05 18

Marc Newson Ltd, 1 Heddon Street, Piccadilly, London W1R 7LE, UK.
T. (0)207 287 9388 F. (0)207 2879347

Nuno Corporation, Axis B1F 5-17-1 Roppongi, Minato-ku, Tokyo, 106-0032, Japan.
T. (0)3 3582 7997 F. (0)3 3589 3439

Nya Nordiska Textiles GmbH, An den Ratweisen, Dannenberg, D-29451, Germany. T. (0)5861 809 43 F. (0)5861 809 12 E. nya@nya.de

Nucleo Global Design Factory, Via Piossasco 29/b, 10152 Torino, Italy. T. (0)011 249 0013
F. (0)011 247 5066 E. nucleo@nucleo.to

OXO International, 75 Ninth Avenue, 5th Floor, 10011 New York, USA. T. (0)212 242 3333

Atelier Satyendra Pakhalé, 70 R.J.H Fortuynplein, 1019 WL Amsterdam, The Netherlands.
T. (0)20 419 72 30 F.(0)20 419 72 31
E. satyen@euronet.nl

Philips Electronics BV, 24 Emmasingel, Eindhoven, 5611 AZ, The Netherlands.
T. (0)40 27 59 006 F. (0)40 27 59 091

Francesco Pineider SpA, Via del Roseto 54, 50144 Firenze, Italy. T. (0)055 62311 F. (0)055 696 390

Privatbank IHAG Zurich AG, Bleicherweg 18, 8022 Zurich, Switzerland.
T. (0)1 205 11 11 F. (0)1 205 12 85
E. info@pbihag.ch

Pro-cord srl, Via del Pratello 9, 40122 Milan, Italy.
T. (0)051 238862 F. (0)051 228368
E. pro-cord@interbusiness.it

Pure Co, 16646 111 Avenue, Edmonton T5M 2S5, Alberta,Canada.
T. (0)1 800 483 5643 F. (0)780 483 5661

Radice SNC, Via Kennedy 22, 22060 Figino Serenza, Como, Italy. T. (0)031 780146
F. (0)031 780289 E. info@radice.it

Regent Beleuchtungskorper AG Basel, Dornacherstrasse 390, CH-4018 Basel, Switzerland.
T. (0)61 335 54 83 F. (0)61 335 55 96
E. export.bs@regent.ch

Rhoss SpA, Viale Aquileia 75, 33170 Pordenone, Italy. T (0)0434 549111 F. (0)0434 43575
E. rhoss@rhoss.it

Roncato SpA, Via Della Pioga 3, 35011 Campodarsego (PD) Italy.
T. (0)049 9200311 F. (0)049 5555570
E. ufficio.tecnico@vroncato.it

Rosenthal AG, Wittelbacherstrasse 43, Selb, D-95100, Germany.
T (0)9287 72586 F. (0)9287 72271

Lorenzo Rubelli SpA, San Marco 3877, 30124 Venice (VE), Italy.
T. (0)041 521 6411 F. (0)41 5225557
E. info@rubelli.it

Santos & Adolfsdóttir, 4 Middle Ground, Favant, Salisbury, Wiltshire SP3 5IP, UK.
T. (0)1722 714669 E. sa-tex@dircon.co.uk

Sawaya & Moroni SpA, Via Andgari 18, Milan 20121, Italy. T. (0)2 86 39 51 F. (0)2 86 46 48 31

sdb Industries BV, De Beverspijken 20, 5221 ED Den Bosch, The Netherlands. T. (0)736339133
F. (0)736312422 E. info@sdb-industries.nl

de Sede AG, Oberes Zelgi 2, CH-5313 Klingnau, Switzerland. T. (0)56 2680 226 F. (0)56 2680 126
E. info@desede.ch

Serafino Zani srl, Via Bosca 24/26, 25066 Lumezzane Gazzolo (BS), Italy. T. (0)30 871861
F. (0)30 89 70 620 E. info@serafinozani.it

Sele 2, Gustav Maurer Strasse 8, 87702 Zollikon, Zurich, Switzerland. T. (0)41 1 396 70 12
F. (0)41 1 396 70 11 E. sele2@sele2.ch

Sharp Corporation, Corporate Design Centre, 22-22 Nigaike-cho, Abeno-ku, Osaka, 545 8522, Japan. T. (0)6 6621 3637 F. (0)6 6629 1162

Siemens Electrogeräte GmbH, Hichstrasse 17, D-81669 Munich, Germany. T (0)89 45 90 29 29/29 37 F. (0)89 45 90 20 31
E. christophe.pott-sudholt@bsgh.com

Snowcrash, Linnegaten 23, Box 483, S-35 06 Vaxjo, Sweden, T/F (0)470 74 24 00
E. info@snowcrash.se

Michael Sodeau, 26 Rosebery Avenue, London EC1R 4SX, UK. T. (0)20 7833 5020
F. (0)20 7833 5021 E. michael@msp.uk.com

Sony Corporation, 6-7-35 Kitashinagawa, Shinagawa –ku, Tokyo 141-0001, Japan,
T. (0)3 5445 6780 F. (0)3 5448 7823

Sony Television Europe, The Heights, Brooklands, Weybridge, Surrey KT13 0XW, UK.
T (0)1932 816190 F. (0)1932 817003

Spatial Interference, 63 Cross Street, London N1 2BB, UK. T. (0)20 7704 6113 F (0)20 7688 0478

Studio Sipek, 7 Jeleni, Prague 11800, Czech Republic. T. (0)424 24373647 F. (0)424 24372331

Spirix Lexon, 98 ier Boulevard Eloise, 95100 Argenteuil, France. T. (0)39 47 04 00
F. (0)39 47 07 59 E. spirix@lexon-design.com

Norma Starszakowna, 102 Royle Building, 23 Wenlock Road, London N1 7SH, UK.
T. (0)20 7251 3873 F. (0)20 7514 7729
E. N.STARSZAKOWNA@LCF.LINST.AC.UK

Stile Bertone SpA, Via Roma 1, 10040 Capri, (Torino) Italy.
T. (0)011 9638 322 F. (0)011 9632 003

Team 7, Naturlich Womnen GmbH, PO Box 228, A-4910 Ried, Austria.
T (0)7752 977 169 F. (0)7752 977 222

Gebruder Thonet GmbH, Postfach 1520 D-35059, Michael Thnet Strade 1, D-35066 Frankenberg, Germany. T. (0)6351 508 0
F. (0)6451 508 108 E. info@thonet.de

TKO Design, 37 Stukeley Street, London WC2B 5LT, UK. T. (0)20 7404 2404 F. (0)20 7404 2405
E. mail@tkodesign.co.uk

Tronconi srl, Via Bernini 5/7, 20094 Corsico (Mi), Italy. T. (0)02 45867089 F. (0)02 4585011
E. tronconi@tronconi.com

Tuttoespresso, 23/25 Via Per Caronna, 21040 Origgio (VA), Italy.
T. (0)02 96730600 F. (0)02 96731856

Umbra, 2358 Midland Avenue, Scarborough, M1S 1P8, Ontario, Canada.
T. (0)800 387 5122 F. (0)416 299 1706

Valcucine SpA, Via Malignani 5, 33170 Pordenone, Italy. T. (0)434 517911 F. (0)434 572344

Dick Van Hoff, Staringplein 14, 6821 DS Arnhem, The Netherlands.
T/F (0)26 4422707 E. vanhoff@knoware.nl

Christine Van der Hurd, 2/17 Chelsea Harbour Design Centre, London SW10 OXE, UK.
T. (0)20 7351 6332 F. (0)20 7376 3574
E. rugs@rugsCVDH.demon.co.uk

Venini SpA, Fondamenta Vetrai 50, 30142 Murano Venezia, Italy
T. (0)417 39955 F. (0)4 17 39 369

Ventura Design, OL Time SA, 8064 Volkeswit, Switzerland. T. (0)1 908 55 99 F. (0)1 908 55 22

Verso Design OY, Mantypaadentie 7 B, FIN-00830 Helsinki, Finland. T/F. (0)358 9 755 47 40
E. verso.design@versodesign.fi

Vitra (International) AG, Klunenfedstrasse 22, ch-127 Birsfelden, Switzerland. T. (0)61 377 15 09
F. (0)61 377 15 10 E. info@vitra.com

VS Vereinigte Spezial Möbelfabriken, 97941 Tauberbischofabriken, Hochauser Strasse 8, Germany. T. (0)09341/880

Norbert Wangen, Grafinger Strasse 6, 81671 Munich, Germany.
T. (0)89 49 001 572 F. (0)89 49 001 573

Carol Westfall, 208–17 West Shearwater Court, Jersey City, NJ 07305, USA. T. (0)201 332 0135
F. (0)201 332 0027 E. delpreor@interactive.net

Wittman Mobelwerkstatten GmbH, A-3492 Etsdorf/Kamp, Austria. T. (0)27 35 /28 71
F. (0)27 35 /28 77 E. wittmann @ wittmann.at

Wogg AG, Im Ginnd 16, 5405 Baden, Aargan, Switzerland. T. (0)56 493 38 21 F. (0)56 493 40 87

WMF, Erberhardstrasse 1, 73312 Geislingen/Steige, Germany.
T. (0)7331 258231 F. (0)7331 258997

XO, RN 19, Servon 77170 Paris, France.
T. (0)1 60 62 60 60 F. (0)1 60 62 60 62

Yamaha Corporation, 10-1 Nakazawa-cho, Hamamatsu, Shizuoka Prefecture, 430 8650, Japan. T. (0)53 460 2883 F. (0)53 4673 4992

Kazuhiro Yamanaka, 25 Woodvale Way, London NW11 8SF, UK. T (0)208 452 3018 F. (0)208 452 3018 E. kaz@ma.kew.net

Zanotta SpA, Via Vittorio Veneto 57, Nova Milanese, (Mi) 20054, Italy. T. (0)362 36 83 30 F. (0)362 45 10 38 E. zanottaspa@zanotta.it

Zeus, c/o Noto srl C, so San Gottardo 21/9, Milan 21036, Italy. T. (0)2 89 40 11 42 F. (0)2 89 40 11 98
E. zeusnoto@tini.it

Zumtobel Staff GmbH, Schweizer Strasse 30, Postfach 72, A- 6851 Dornbirn, Austria. T. (0)5572 390 0 F. (0)5572 20 721
E. info@zumtobelstaff.co.at

Photographic credits

The publisher and editors would like to thank the designers and manufacturers who submitted work for inclusion, and the following photographers and copyright holders for the use of their material (page numbers are given in parentheses):

Riccardo Abbondanca (200 top centre)
Maos Armgaard (187 bottom right)
Toshi Asakawa (55)
Clive Bartlett (153 bottom)
Rien Bazen (133)
Jan Benatsson (200 bottom right)
Joachim Bergamin (79 top)
Fabrizio Bergamo (47 top)
Francesco Bighin (90 top left)
Bitetto-Chimenti (101 left, 102 left, 111 left)
Bius (27 bottom)
BMW GB Ltd (222)
Allessandro Bon (198 top right, bottom right)
Todd Bracher (24 bottom left)
Erik Brahl (22–3)
Brahl Fotografi (57)
Bruno Bruchi (132 top, 148, 154)
Bulthaup (217)
Santi Caleca (112 top right, 143, 153 top left, top right)
Cinzia Camela, Sara de Bernardinis (25 bottom)
Pietro Carrieri (32 bottom)
Giampetro Casadei (36 top left, top centre, top right)
Armen Casnati (83 top left, bottom left)
Scott Chaney (176 top left, bottom)
Corado Maria Crisciani (105)
Geoff Crowther (141)
Karsten Damstadt (131, 201)
Design 3 (186 bottom right)
D. James Dee (175)
Hans Doering (124 top right)
Eastman Kodak Company (181 bottom left)
Claus-Christian Eckhardt (186 top right, centre right)
Rick English (189 bottom centre, 190 top right, bottom right, 210 top left, centre left, middle left, middle centre, borrom left)
EOS (78)
Estudycolor (112 bottom)
Fabrice für Fotografi (200 top left)
Ramak Fazel (79 bottom left, bottom right, 80 bottom left, 120)
D. Feintrenie (29 right)
Frans Feijen (157)
Christophe Fillioux (43 top right, bottom right)
Facchinetti Foriani (92, 93 top left)
Friedrich Forssmann (60 top left)
Lyn Gardiner (98 right, 99)
Hynek Glos (149 top right, bottom right)
Enrico Graglia (168 top left)
© Tobias Grau/Michael Wurzbach (90 right, bottom left, 124 top left, centre left)
Rolando Paolo Guerzoni (210 bottom right)
Walter Gumiero (38 top left, top right, bottom left, bottom right, 39–41)
Hans Hansen (33 bottom, top right)
Patrick Hanssens (114–5)
Hiroki Hayashi (193 top left, top right, bottom left)
Hayo Heye/Stefan Thurmann (63)
Claudy Jongstra (164)
Bernd Kammerer (18 top right, bottom right)
Lasse Keitto (174 top right)
Malcolm Kennard (221 top left, bottom left)
Jutta Kennepohl (64 top right, bottom right)
Tobias Koeppe (205 right)
Iveta Kopicova (80 top left, top right)
Rene Koster (35 bottom left)
Richard Learoyd (171 top left, bottom left)

Bernhard Lehn (94 left)
K.C. Lo (129)
Lars Mardahl (71)
Ramazzotti Marino (49 top right, bottom right, bottom left)
Kenny McCracken (142 top left, top right)
Ian McKinnell (186 bottom left, 211)
Jaime Miró (111 top right, bottom right)
Ricardo Moncada (25 top, 137)
Carlo Monzio/Gianmarco Bassi (110 top right, bottom right)
O. Moritz (180, 186 top left)
Daniel Nicolas (34 top centre, centre right, 35 top, bottom right)
Søren Nielsen (33 top centre, 42 top left, bottom left)
Yoshi Nishima (155)
Manuel Nunes (34 top right)
Oxo International (218 top right)
A. Paderni (46 top right)
Pallucco Italia SpA (123 top left)
Keith Parry (30 bottom left)
Paterno (77)
Antje Peters (221 right)
Bianca Pilet (84–5)
Pink Moon (26 top centre, 45 top left, 72 bottom left)
Carlos Piratininga (103)
Walter Posern (93 top right)
Amiel Pretsch/AS (48 left)
Miguel Ribot (124 bottom left)
Francesco Riva/Smoky Minds (134)
Romano Fotografie (122 right)
Jonathan Rose (18 bottom left)
Finn Rosted (136 bottom)
Ilan Rubin (18 top left)
Hideto Sasaki (207)
Pietro Savorelli/Studio 33 (212 top right, bottom right)
R. Schmutz (72 bottom centre, bottom right)
Frank Schwarzbach (208)
Jurgen Schwopp (140 bottom right)
Luigi Sciuccati (194 top left, bottom left)
Altin Sezer (162–3)
Filip Slapal (27 top right, 28 top right, bottom right, 74 top left, bottom, 130 bottom right, 142 bottom, 149 left)
Sharp Corporation Advertising Division (182–3, 192 top right, 198 left, 199)
Luciano Soave (82 top right)
David Spero (94 centre right)
David Steets (104 top right)
Thomas Stewart (19 bottom left)
Studio Bitetto-Chimenti (47 bottom left, bottom right, 214, 215 bottom)
Studio Bertone SpA (223 bottom)
Studio Feruglio (26 top right, bottom right)
Studio Fuoco (98 left)
Studio Haas (218 top left)
Studio Reflex (44 top left)
Studio Synthesis (28 top left)
Studio Leo Torri (218 bottom left, bottom centre)
Studio Uno (68 top left, centre left, bottom left)
Peter Tahl (16–17)
Kozo Takayama (107)
Luca Tamburlini (26 top left, centre left, bottom left, 36 bottom left, bottom right, 187 bottom left)
Andy Taylor (165)
Alexander Tsoehler (65)

Paolo Ulian (204)
Tom Vack (51, 96–7)
© Patricia von Ah (14–15, 21)
Reinout van den Bergh (34 bottom left, bottom centre, right, top left)
Toine van den Nieuwendijk (14 bottom left)
Camille Vivier (220)
Pelle Wahlgren (33 top left)
Felix Wey (76, 89, 123 top right, bottom right)
Gianluca Widmer (145 top right, bottom right, 93 bottom left)
Jonata Xerra (106)
Miro Zagnoli (88 right, 116, 121 left)
Max Zambelli (104 top left)
David Zanardi (210 bottom middle)

240

Learning Resources Service

Somerset College of Arts and Technology
Wellington Road Taunton TA1 5AX
T 01823 366 469

SOMERSET COLLEGES

T0093400

This book is due for return on or before the last date shown below

1 2 FEB 2002	0 9 MAR 2004
1 8 FEB 2002	2 3 MAR 2004
-7 MAR 2002	1 1 FEB 2005
15. APR 02.	-1 MAR 2007
1 6 APR 2002	CANCELLED 02 APR
	-7 MAR 2007
1 6 MAY 2002	
-7 JUN 2002	
1 3 JAN 2003	
2 9 SEP 2003	